Fatigue in railway infrastructure

Related titles:

Wheel–rail interface handbook
(ISBN 978-1-84569-412-8)
Many of the engineering problems of particular importance to railways arise at interfaces and the safety-critical role of the wheel–rail interface is widely acknowledged. Better understanding of wheel–rail interfaces is therefore critical to improving the capacity, reliability and safety of the railway system. *Wheel–rail interface handbook* is a one-stop reference for railway engineering practitioners and academic researchers. Part I provides the fundamentals of contact mechanics, wear, fatigue and lubrication as well as state-of-the-art research and emerging technologies related to the wheel–rail interface and its management. Part II offers an overview of industrial practice from several different regions of the world, thereby providing an invaluable international perspective with practitioners' experience of managing the wheel–rail interface in a variety of environments and circumstances.

Inspection and monitoring techniques for bridges and civil structures
(ISBN 978-1-85573-939-0)
This important reference work reviews ways of detecting corrosion in steel and concrete used in bridges as well as decay in timber components. It also discusses monitoring techniques such as acoustic emission and electrical impedance testing.

Fatigue in railway infrastructure

Edited by
Mark Robinson and Ajay Kapoor

CRC Press
Boca Raton Boston New York Washington, DC

WOODHEAD PUBLISHING LIMITED
Oxford Cambridge New Delhi

Published by Woodhead Publishing Limited, Abington Hall, Granta Park
Great Abington, Cambridge CB21 6AH, UK
www.woodheadpublishing.com

Woodhead Publishing India Private Limited, G-2, Vardaan House, 7/28 Ansari Road, Daryaganj, New Delhi – 110002, India

Published in North America by CRC Press LLC, 6000 Broken Sound Parkway, NW, Suite 300, Boca Raton, FL 33487, USA

First published 2009, Woodhead Publishing Limited and CRC Press LLC

British Library Cataloguing in Publication Data
A catalogue record for this book is available from the British Library.

Library of Congress Cataloging in Publication Data
A catalog record for this book is available from the Library of Congress.

Woodhead Publishing Limited ISBN 978-1-85573-740-2 (book)
Woodhead Publishing Limited ISBN 978-1-84569-702-0 (e-book)
CRC Press ISBN 978-1-4398-1873-2
CRC Press order number: N10123

The publishers' policy is to use permanent paper from mills that operate a sustainable forestry policy, and which has been manufactured from pulp which is processed using acid-free and elemental chlorine-free practices. Furthermore, the publishers ensure that the text paper and cover board used have met acceptable environmental accreditation standards.

Typeset by Godiva Publishing Services Limited, Coventry, West Midlands, UK
Printed by TJ International Limited, Padstow, Cornwall, UK

Contents

Contributor contact details

(* = main contact)

Editors
Professor Mark Robinson
School of Mechanical and Systems Engineering
Stephenson Building
University of Newcastle
Newcastle-upon-Tyne NE1 7RU
UK
E-mail: newrail@ncl.ac.uk

Professor Ajay Kapoor
Faculty of Engineering
Swinburne University of Technology
PO Box 218
Hawthorn
Victoria 3122
Australia
E-mail: akapoor@swin.edu.au

Chapter 1
Professor R. A. Smith
Future Rail Research Centre
Department of Mechanical Engineering
Imperial College London
London SW7 2BX
UK
E-mail: roderick.smith@imperial.ac.uk

Chapter 2
Professor L. Lesley
30 Moss Lane
Orrell Park
Liverpool L9 8AJ
UK
E-mail: ljlesley@aol.com

Chapter 3
Dr M. Gilbert
Department of Civil and Structural Engineering
The University of Sheffield
Sir Frederick Mappin Building
Mappin Street
Sheffield S1 3JD
UK
E-mail: m.gilbert@sheffield.ac.uk

Chapter 4
Dipl.-Ing. K. Behrens
E-mail: behrens.karsten@t-online.de

Chapter 5
H.-P. Kohlbecker
Heinrich Klee Str. 12
56294 Münstermaifeld
Germany
E-mail: hans-peter.kohlbecker@bahn.de

Preface

Transport is an important part of the economy. Estimates suggest, for example, that the transport sector represents roughly a tenth of the European Union economy. Additionally, substantial societal benefits accrue in terms of mobility of the population for social and recreational purposes. Economic growth needs good transport infrastructure to move workforce and goods around, and congestion is known to cause damage in billions of dollars in waiting times, lost productivity and wellness of the population.

Urbanisation and the ageing population are putting pressure on public transport, and this pressure is expected to continue to rise in the foreseeable future. With roads predicted to get further clogged and traffic further slowed down, the demand on railways will continue to grow. The green credentials of the railway system, aided by the mass transit capacity (for people and minerals) and the relatively low power needed given the low friction of the steel wheel on steel rail, will become even more important in the coming decades, pushing up demand even further. Railways are going to remain a major form of transport in the future.

Railways are important for another reason as well – the dog and man syndrome within the mass media. We all know that a dog biting a man is not a good story, but a man biting a dog is! The railway's green credentials, its mass transit capacity, and its ability to provide speedy access to towns and cities seldom make news in the public media. Accidents do, and rightly so. Recent train accidents have focused public opinion and significant resources on the causes of failure in wheels (the June 1998 accident in Eschede, Germany), rails (the October 2000 accident at Hatfield in the UK), points (the May 2002 accident at Potters Bar, UK), and signals (the September 2008 accident in the San Fernando Valley in southern California, USA) to name just a few.

Several of these recent and not so recent accidents were caused by fatigue. Fatigue in the railway system has been studied for a long time and many issues are well understood, but the push for lighter vehicles, higher speeds and more comfortable journeys provides newer challenges every day. In order to provide increased capacity in terms of both passengers and freight, the rail sector is moving towards higher train speeds and increased payloads which lead to larger axle loads. These two factors result in larger forces acting on the railway wheels and tracks which could result in more rapid fatigue behaviour. Research efforts are therefore directed at reducing vehicle weights and improving the wheel–rail interface through better track and kinder bogies.

The area of fatigue is important as it often results in catastrophic failure with little or no warning. These failures can affect all aspects of the rail system but these have the most severe consequences when they occur at the wheel–rail interface: wheels, rail, switches and crossings. Failures are extremely costly in terms of lost revenue, operational delays and human injuries. It is therefore important that rail researchers improve their understanding of rail fatigue and failure mechanisms.

The purpose of this edited volume is to assemble noted experts in the area and collect their views in one place. The book will help in understanding of fatigue in the railway system, with a view to eliminate or at least reduce accidents. Another important issue is the need for maintenance-free railways or, at least, reduced-maintenance railways able to provide a continuous service. This is possible if the time of failure of a component can be predicted accurately. The expert views presented in this book will play a part in developing such a predictive capability.

Thanks are due to the editorial team at Woodhead Publishing, Sheril Leich and Francis Dodds, who worked extremely hard in keeping the project going, a task made more difficult by Ajay's move from UK to Australia.

Ajay Kapoor and Mark Robinson

1

Fatigue and the railways: an overview

R. A. SMITH, Imperial College London, UK

Abstract: This chapter provides an overview of fatigue issues affecting railways. It starts by considering fatigue at the wheel–rail interface, particularly the effect of dynamic loads on rails, bearings, axles, suspension and other components. It then reviews fatigue issues in vehicles such as body shells, engines, couplings and internal components. Finally, it considers fatigue in infrastructure such as bridges and signalling systems.

Key words: fatigue in railways, wheel–rail interface, dynamic loads, rail vehicles, rail infrastructure

1.1 Introduction

Study of the type of material failure we now call fatigue originated from problems on the railways about 200 years ago. Fatigue still causes failures in many kinds of components, in many different industries, and failures still occur in railways. In recent years, two accidents caused by fatigue had particularly serious repercussions. On 3 June 1998, a German Intercity Express (ICE) was derailed after a wheel failed and then subsequently caught in a set of points, causing a carriage to strike the support of an over-bridge. The bridge collapsed and several carriages piled up in the debris of the bridge; 101 people were killed. A recent paper gives some background, but is incomplete in detail (Esslinger *et al.*, 2004). On 17 October 2000, a British train derailed at Hatfield, just north of London, killing four passengers. The immediate cause of the derailment was identified as a broken rail, and a subsequent examination of the UK network led to the discovery of more than 2000 sites containing potentially dangerous cracks. Severe speed restrictions were imposed while repair and replacement of track took place over a period of many months. In the long history of Britain's railways, no previous accident had caused such widespread public anger, managerial panic, disruption and eventual political crisis (Jack, 2001; Murray, 2001; Wolmar, 2001). The railway system had been privatised between 1996 and 1998, by fragmenting it into more than 125 companies and separating operations from infrastructure, the latter being a common feature of several other privatisations in other countries. As a consequence of the Hatfield accident and its aftermath, Railtrack, the UK infrastructure company, was taken into receivership in October 2001 and was subsequently reformed as a 'not-for-profit' company, Network Rail. More recently, changes in the organisational structure of the railway designed to reduce fragmentation have been announced (Anon., 2004).

At first sight it seems surprising that, despite its long history, catastrophic fatigue failures still occur. But these well-publicised accidents are only the tip of the iceberg. The consequences of a failure depend on a chance chain of events occurring after the failure. In most cases, usually by good fortune, the consequences are not so severe, but much can still be learnt from all incidents, so that, by investigation and good reporting, future similar occurrences may be reduced.

It is the purpose of this chapter to provide a broad overview of fatigue failures in railways and to make some suggestions about the most effective ways of reducing their incidence. The fatigue process itself will not be discussed in detail: many excellent books and reviews are readily available (Suresh, 1998; Schijve, 2001, 2003).

It is probably worth noting that the term fatigue here is used throughout to mean the fatigue of materials and not human fatigue due to tiredness. This latter kind of fatigue also has a long history of causing accidents on railways!

1.2 Fatigue and railways

Railways are characterised by the contact between the vehicle's wheels and the rails, the guidance system. The major benefit of railways as a transportation system stems from this key feature. Because this contact is very stiff, the rolling resistance is low, so that heavy loads can be hauled with comparatively small tractive effort. Indeed, much of the operation of a railway system is determined by the peculiarities of the wheel–rail interface. Acceleration and braking are determined by adhesion, stopping distances define the characteristics of the signalling and control system and the latter defines the capacity of the system. It will be noted that both the major failures referred to above were at the wheel–rail interface, and it is failures at this location that are particular to railways. There are, however, a large number of other fatigue failures that concern the railways, but they are not necessarily railway specific. Table 1.1 attempts to summarise most of the major areas of concern and defines the order in which the discussion that forms the remainder of this chapter is structured.

It should be recognised that railway equipment operates in a hostile environment, which is often dirty and wet. Perhaps surprisingly, despite long experience, the loads and stresses to which equipment is subjected are often inadequately defined. This is particularly true of dynamically induced loads of which more is said later. Rolling stock is designed to last for between 30 and 40 years but is often used for much longer. Major infrastructure can last for considerably more than 100 years. These very long service lives mean that the obvious principal design requirement against fatigue failure is that stress ranges

Table 1.1 Significant areas of fatigue in railways

Adjacent to wheel–rail interface	Wheels Rails Rail welds
Affected by forces generated at the wheel–rail interface	Bearings Axles Gearboxes Drive shafts Bogies Springs and suspension components Brake components Rail fastenings and supports Track foundation
Vehicles	Engine or motor components Body shells Couplings Internal components and fittings
Infrastructure	Bridges Signals Electrical supply components

should be below the fatigue limit. This apparently simple requirement is not as easy to apply as may be imagined, partly because the loading spectrum can contain many larger load excursions superimposed on a baseline of constant amplitude loading, and partly because of competing deterioration mechanisms such as wear and corrosion which can erode the original design margin. The fatigue limit concept was determined from extensive experiments conducted by the German railway engineer Wöhler, in a study of the fatigue of axles, Fig. 1.1 (Wöhler, 1858–1871). Despite its long use, there is growing evidence that for lives longer than the conventional 10^6–10^7 cycles at which the fatigue limit is determined, the safe stress range continues to be eroded down to 10^9 cycles and more, that is, at the very long lives typical of that required of axles and wheels (Stanzel-Tschegg, 2002).

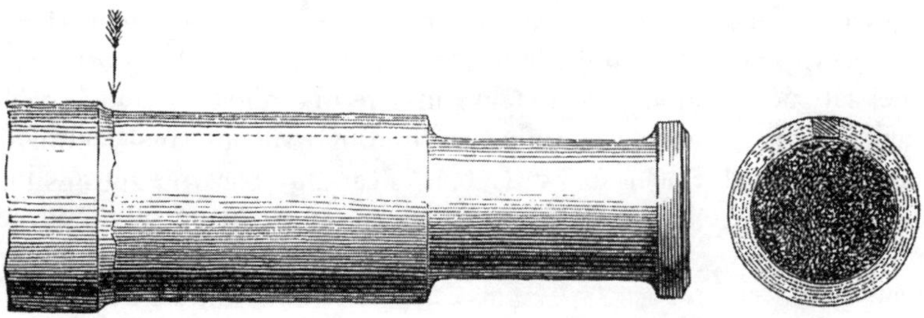

1.1 A typical axle failure of the 1840s (shown by arrow) leading to the birth of the fatigue problem.

Although, in general, it may be said that most fatigue problems have been satisfactorily solved, the need for safety means that components must be subjected to expensive inspections in order to guard against a small number of possible failures. These inspections are not always reliable in identifying possible deterioration, and the dismantling necessary to achieve access can often introduce inadvertent damage.

The long service lives of railway equipment mean that the 'technological window' for railways is particularly wide. New technologies take a long time to be implemented across the whole system and must work side by side with existing equipment during the substitution period. For example, a new improved rail steel, however great its advantages, will not be in use system-wide for at least 50–60 years. Again, improved information technology, easily applied to new-build vehicles, may only be applicable to existing vehicles by expensive retrofits. By contrast, the automobile industry renews itself almost completely in a 10-year life cycle. These issues will be amplified by discussing the various areas of fatigue shown in Table 1.1.

1.3 Fatigue at the wheel–rail interface

A single contact patch between the wheel and the rail is typically the size of a small coin: a long train is completely supported over a total area no larger than that of a compact disc. Clearly, the pressures at the key interface are very high, considerably in excess of the normal yield stress of the material. A complex series of events takes place with repeated passages of a wheel over a rail. The material in the immediate vicinity of the contact work-hardens and deforms until its 'ductility is exhausted'* and a series of small cracks forms. Ideally, if the wear rate of the railhead or wheel equals or exceeds the rate at which cracks are initiated, then the cracks are 'rubbed out' before they can develop. However, if the crack development rate exceeds the wear rate, the cracks propagate deeper into the material, driven by the contact stresses. As the contact stresses diminish with depth into the material, the bulk stresses in the interior of the wheel or rail take over as the drivers of the crack. The possibility therefore exists of non-propagating cracks, if 'handshakes' fail to happen in the zones of transfer in the sequence of the change-over of the governing stress from the surface stress to the contact zone stress to the bulk stress. This type of behaviour is paralleled in other fatigue situations when cracks initiate in high surface stress fields at, for example, sharp geometric notches, fretting patches and thermally loaded surfaces. In both wheels and rails, cracks can turn back up towards the surface, leading to the formation of a detached flake (spalling).

1. This somewhat old-fashioned term means that the yield stress of the material is raised to some limiting value by the repeated plastic deformation in the contact zone. The process is referred to as ratchetting.

1.3.1 Fatigue of wheels

This kind of spalling damage is relatively common on railway wheels. It leads to poor running conditions and high dynamic impact loads. In most cases this damage, if caught in its early stages, can be removed by re-turning the tread of the wheel. Similarly, out-of-roundness (polygonisation) or wheel flats, caused by sliding, can be machined out before damage becomes too widespread. Turning is used in the first instance to re-profile the wheel, in order to improve contact patch conditions, which are particularly sensitive to the local geometries of the wheel and rail at the site of the contact. In the past, wheels were usually manufactured by shrink fitting a tyre onto a hub. The famous 'wheel tappers', whom older readers may remember, were listening principally for loose tyres rather than for cracks as is often supposed. Modern practice is to make wheels of a monobloc construction, with a relatively thin web, curved in the plane of the wheel to give lateral strength through geometry. Failures in the web are rare. However, despite all our knowledge of stress concentrations, a recent wheel fracture on a high-speed train initiated at a hole that had been drilled into the web of the wheel in order to attach a balance weight. The wheel disintegrated, but the train was fortunately able to come to a halt without causing any casualties (a good example of fate being kind, and the failure not unleashing a catastrophic series of events). This obviously dangerous method of balancing has been ceased. The wheels are now balanced by eccentric machining of the interior underside of the rim in a manner which achieves balance by removing a small crescent-shaped piece of material smoothly blended into the profile, thus avoiding any stress concentrating discontinuities.

The much-publicised accident to the German ICE train in 1998, which resulted in more than 100 fatalities, was caused by a fatigue fracture on the underside of a rimmed wheel separated from the disc of the wheel by rubber pads. This design, much used on vehicles operating at lower speeds, has the supposed advantage of reducing the transmission of noise and vibration from the wheel–rail contact into the body of the vehicle. The so-called resilient wheels were put into service without, in the author's opinion of the evidence available to him, adequate fatigue testing. In particular the amount of material that could safely be removed from the tread to re-profile the wheel was not determined. As more and more material was removed in successive turning operations, the tyre became, in effect, a more flexible thinner ring. The squeezing of this ring caused by the rotation of the wheel led to high bending stresses on the inside of the tyre. The inspection techniques were concentrated on the outer tread of the wheel, the usual site of contact fatigue damage in a solid wheel. It appears that the inadequate testing had not been continued to produce failure, thus the site of potential failure was unknown and not adequately covered by the inspections.

1.3.2 Fatigue of rails

The history of rail failures is as long as that of the railways. Cast iron was replaced by wrought iron, before itself being superseded by steel from 1860 onwards. In the last 30 years, the quality of steel manufacture has improved, virtually eliminating fatigue failures initiated from internal inclusion or hydrogen shrinkage defects in the railhead. Probably the most significant development since the introduction of the steel rail has been the use of welding to eliminate fish-plated gaps in the running surface. Rail is now manufactured in strings up to 250 m long, thus simplifying the laying of track. The weld is itself a source of potential weakness: a large proportion of rail failures now occur at these joints. The thermit welding process is used in the field to join long rail strings. This process uses the exothermic reaction of a mixture of iron oxide and aluminium powder to connect the rail ends by what is essentially a casting. Techniques are continuously being improved (Mutton and Alvarez, 2004), but quality control under often adverse conditions is difficult and it is no surprise that defective welds are impossible to eliminate completely. Inspection techniques for welds have also improved, but are still not infallible. For example, there are currently over 130 000 welds installed in the UK railway infrastructure each year and it is estimated that there are in excess of 2.5 million in the track. These very large numbers serve to emphasise the potential dangers caused by even an extremely low percentage failure rate.

The life of a rail is principally determined by wear at the railhead. This wear can, in certain circumstances, produce a short-wavelength shape change along the length of the rail, known as corrugation, which in turn leads to poor ride and noise generation. Controlled grinding is used to remove corrugations and/or to restore the accurate lateral railhead profiles that are essential for controlling the stresses in the wheel–rail contact. Combinations of high contact stresses upon which traction stress (along the rail) or cornering stresses (across the rail) are superimposed can lead to the initiation of rolling contact fatigue cracks (Smith, 2001). The particular types of these cracks caused by cornering are situated to the inside of the railhead and are known as 'gauge-corner' cracks. If the wear rate is greater than the rate of development of fatigue cracks, the deterioration of the rail is benign. If, however, wear rates are low, it is possible for fatigue cracks to grow down into the railhead. The cracks progress at a slow rate from the running surface, typically inclined downwards at a shallow angle of some 10°, until some 5 mm below the surface they branch. If the branch crack propagates upwards, driven by plastic deformation of the thin tongue of metal above the crack, a part of the rail surface detaches or spalls – a form of damage that is clearly visible on inspection. But more dangerously, some cracks turn downwards into the head of the rail and these branches are extremely difficult to detect by conventional ultrasonic inspection techniques. It is claimed that eddy current methods may be more reliable, but experience in the field is so far limited. If cracks remain undetected they can eventually grow in the zone of influence of the gross bending stresses in the body of the rail, turn downwards, propagate across the cross-section of the rail and eventually become large enough to cause complete fracture of the rail. This mechanism was in fact the cause of the rail fracture at Hatfield mentioned at the beginning of this chapter.

It is tempting to reduce wear rates on heavily trafficked sections of track by increasing the surface hardness of the rail. This action may, however, tip the balance to make fatigue crack propagation the dominant mode of deterioration. Subsequent grinding might then not remove the propagating sub-surface tips of the cracks. A great deal of work – theoretical, laboratory based and derived from experiments in service – has been performed on this problem over the last two decades. There is now sufficient knowledge available to control this potentially dangerous problem, by a combination of inspections, grinding and contact stress reduction. The problem is such that many parameters, involving both the rail and the vehicle (wheel profile, suspension characteristics, etc.), need careful consideration. In railway systems where responsibility for the track and the vehicle has been placed with different authorities, care is needed to ensure that there exist mechanisms for those in charge of both sides of the wheel–rail interface to understand the complexities of the problem and to act in unison. The topic of fatigue at the wheel–rail interface, with particular focus on the rail, was extensively covered in a recent special issue of a specialist fatigue journal (Lundén, 2003). In this issue, after three scene-setting review articles, there follow 12 research articles, three on monitoring, maintenance and non-destructive testing, four on damage, fatigue and fracture of rails, three on phenomena at the wheel–rail interface and, finally, two on new rail materials. The attention of readers is particularly drawn to this recent overview of technical, scientific and practical aspects of fatigue at the wheel–rail interface.

1.4 Fatigue affected by forces generated at the wheel–rail interface: the importance of dynamic loads

We now turn from the wheel and rail, components obviously and directly affected by the stresses generated at the wheel–rail contact, to components away from the vicinity of the contact but nevertheless affected by the conditions at the contact. It is worth pausing to mention the nature of the forces at the contact. At its simplest level, the contact patch between each wheel and the rail must support that proportion of the vertical static load, the weight, which passes through it. Because of symmetry, this is known as the axle load (the wheel load equals half the axle load). In addition, along the direction of the rail, forces due to the acceleration, braking and traction at steady speed must be sustained. When a train passes through a curve, the lateral loads needed to generate curved motion must be considered, together with the load redistribution from inner (lower) to outer (higher) rail. All these loads are relatively easy to quantify, but the situation is made much more complicated by the generation of dynamic loads.

In a useful review (Hill and Everitt, 1988), some of the historical gropings towards an understanding of this important effect were outlined. Their observations were so pertinent, they are worth quoting at some length:

> Over the years a number of individuals have had the insight to perceive the importance of an appreciation of the service requirements. For example (Beaumont, 1879) observed in 1879 that:
>
> > When a train was running, the wheels were lifted up and down again on the very many irregularities of the line at a velocity which induced severe shocks. The velocity at which impact shocks were transmitted through the wheels to the axle was not simply that of gravity and that of the velocity of the train, but very many of the shocks were thus transmitted at the velocity of recoil of a loaded spring, which was probably as much as 1300 feet per second [$400\,m\,s^{-1}$].
>
> Notwithstanding this observation, the railway axle soon entered folklore as something to be designed with the nominal stress under a fatigue limit (e.g. Anon., 1920).
>
> Between the two world wars fatigue was studied almost exclusively as an endurance limit problem. The attitude is still prevalent today despite publications by people who have actually measured operating loads and strains. For example, a paper from the mid-1950s (Moreau and Peterson, 1955) . . . with remarkable insight on the field testing of diesel locomotive axles. They commented:
>
> > It is now possible to predict, with reasonable accuracy, what stresses will be induced in a specific axle design by a certain load and the relationship between the stress and the number of stress applications which will cause a failure is also fairly well established. . . . There is, however, very little information available about the loads an axle is actually exposed to in service. To determine whether a part of a structure of a machine is strong enough, the engineer must know the type of loading to which it will be exposed. If he does not, he has no other choice than to make a guess and see if it fails. It is too expensive to learn about weaknesses in axles from failures and it is also too expensive to make them so heavy that they are bound to be strong enough in spite of the designer's ignorance about the loads. . . . To study axles under service conditions it is necessary to study axles in high speed passenger service and in slow freight service, on curves and on tangent track, on good track and on bad. The axles might be damaged under conditions which occur only occasionally. To make sure that no such conditions are overlooked, the behaviour of the axles must be observed over long periods.
>
> Over 70 years after Beaumont, Moreau and Peterson found the service operating environment to be very different from the view held by the majority at that time. For example, in the course of their investigation they

observed that about once every 1000 miles [1600 km] a stress of nearly four times the normal value occurred.

One of the major reasons for this state of affairs until about 1940 was the lack of suitable transducers to make the measurements. Until that time, with the exception of a brief period of use of magnetic strain gauges and carbon strip gauges, only mechanical means had been available. These devices were direct descendants of the method reported by the 1849 Commission into the Application of Iron to Railway Structures (Anon., 1849):

> . . . and a pencil was fixed to the underside of one of the girders of the bridge, so that when the latter was deflected by the weight of the engine or train either placed at rest or passing over it, the pencil traced the extent of the deflection upon a drawing board attached to the scaffold.

It is worth discussing these dynamic loads in more detail. It is now recognised that the magnitude of the dynamic loads induced by the passage of a wheel over a discontinuity in the rail, for example, a gap, dip, or damage patch, is determined by, of course, the magnitude of the discontinuity and by the axle load in combination with the 'unsprung mass' of the vehicle, that is, the mass below the main suspension in 'hard' contact with the rail.

1.4.1 An illustration of the magnitude and effect of dynamic loads

An example calculated using a simple model from data supplied by the Japanese Central Railway Company is shown in Fig. 1.2. This figure illustrates the forces generated as a function of time by the passage of a train over a small (5 mm) dip in the rail head. Two trains are shown, an old type (Series 100) and its replacement (Series 300). The intention was to increase the speed of operation from 180 to 230 km/h. The form of the response from both trains at both speeds is similar, with the dynamic forces showing two clear peaks with time, the so-called P_1 and P_2 forces. The dynamic magnification increases with speed and lies in a range approximately 2.5 to 3 times greater than the static force. Clearly, these magnified forces have a significant effect on the fatigue of wheels, rails and axles. They are significant too in their effect on track maintenance. This is summarised in Fig. 1.3, which is a representation of the typical track maintenance costs as a function of speed for both types of train. The important characteristics of the new train are shown: a smaller wheel load (reduced from 7.5 to 5.7 tonnes) and a smaller unsprung mass (reduced from 2.3 to 1.7 tonnes), the reduction of which is a particularly sensitive way of reducing dynamic track forces. In the example shown, if the old train had been run at the required higher speed of 230 km/h, the track maintenance costs would have increased by some 20%. However, the new lighter train produces a saving of some 10% even at the higher speed. Obviously this is a somewhat simplified view of a complex situation which depends on many parameters. However, it serves to capture the essence of

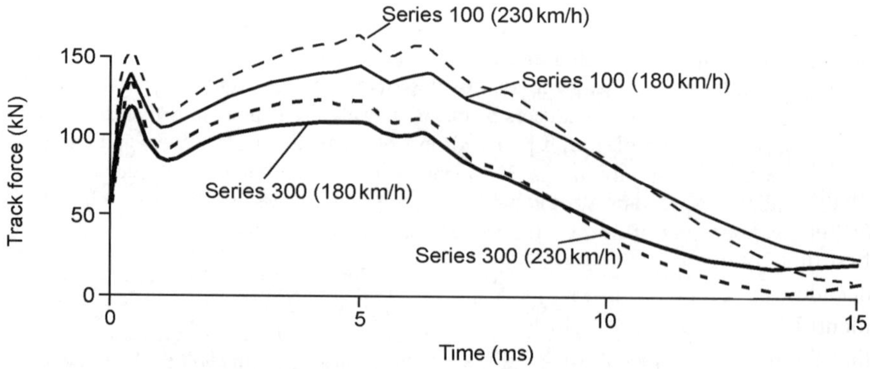

1.2 Dynamic forces produced by the passage of trains over a rail head geometry defect (track force response for a 0.0025 rad, 5 mm rail dip).

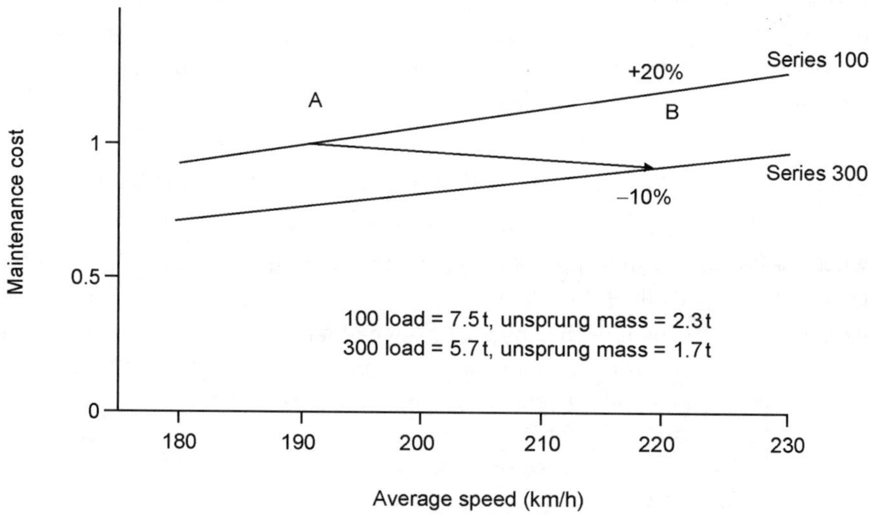

1.3 Generic effect of dynamic forces on maintenance costs.

the dynamic load problem and illustrates the need for track and train designers to work in conjunction with each other. It serves too to illustrate the constraint of higher speeds and structural integrity. For high speeds it is necessary to drive down mass in critical components, thus making them more prone to fatigue.

1.4.2 Bearings and axles

The life of bearings has been much improved by increasing cleanliness of steels. If care is taken to lubricate the bearings correctly and to prevent the entry of dirt, then satisfactory long lives can be easily obtained. There are reports that bearings have failed after dismantling in order to gain access to axles for crack detection examinations. The reason for the need to examine axles arises because of their safety-critical nature and the few, but persistent, number of axles which fail in service.

The fatigue failure of axles in the first railways was the catalyst that led to the identification of fatigue as a failure mechanism. Many investigations were prompted by the accident on the Paris to Versailles railway in 1842, Fig. 1.4, the first time a railway accident had caused major loss of life (Smith, 1990). In later decades, the pioneering experiments of the German engineer Wöhler led to the identification of the fatigue limit for steels. It is something of a surprise therefore that failures still occur. Although it might be assumed that the simplest solution would be to increase the size of axles to reduce stresses, the counter-argument outlined previously is that axles form part of the unsprung mass of the vehicle which must be minimised to reduce the generation of dynamic stresses. Particularly as operational speeds of trains have been increased, the pressure to reduce unsprung mass has become more pressing.

The need for thorough understanding of the service loads to which axles are subjected in service has already been noted. The loading is principally sinusoidal due to the bending couples produced by the upward wheel load reactions being offset from the downward supporting bearing loads. However, the wheel loads can be greatly magnified by dynamic effects, and the equally distributed static loads on each wheel on straight track can be redistributed by cornering and wheel nip at tight gauges as well as by breaking and accelerating forces. The condition of the track and the wheels is paramount in determining the levels of dynamic forces involved, which also increase with speed.

1.4 The Versailles accident of 1842, caused by a broken axle: the first railway accident leading to a large loss of life.

Over the years many experiments have been performed to measure stresses on axles in service. Until recently, the usual procedure was to use slip rings to carry strain gauge signals from the rotating axle into appropriate recording equipment. Because of the bulkiness of the equipment involved, records have been obtained over relatively short times and therefore distances. New developments in electronics have produced miniature equipment containing great recording and processing power. A programme now underway (Smith and Hillmansen, 2001) uses such equipment, which can be directly mounted on the axle and left unattended for periods of several months. A continuous load spectrum is recorded for later analysis by, for example, rain flow counting. More interestingly, strain is recorded over short time intervals of about five seconds, but only retained if a large strain event triggers a storage command, together with a location signal via a Global Positioning Satellite signal. It is hoped that the key very large strain excursions will be captured and identified in this way, in order to clarify why failures occur and how the severity of loading is related to track condition.

1.4.3 Inspection of axles and crack detection in axles

Although failures of axles are rare, typically two or three per year on the UK railway system, the consequences can be catastrophic. Therefore great effort and cost are expended in examining axles for cracks based on a philosophy of setting inspection intervals which leave some margin in the time it would take for the largest non-detectable crack to grow to failure in the time between inspections. However, because of the relatively large sizes of cracks which may be reliably detected (orders of several mm) and the runaway nature of fatigue crack growth, see Fig. 1.5, it is not easy to set economic yet practical inspection

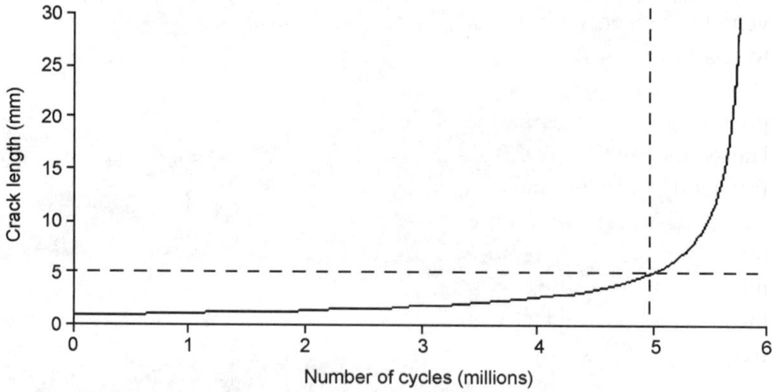

1.5 Crack length plotted as a function of number of cycles. The initial defect size is chosen to be 100 μm and axle failure is assumed to follow rapidly after the crack has grown to 30 mm. A crack length of 5 mm can be detected with a reasonable degree of certainty using NDT methods. This figure clearly illustrates that once a crack length of 5 mm is attained, the axle is near the end of its life.

intervals. Non-destructive testing methods, ranging from ultrasonics, magnetic particles, dye-penetrants and eddy currents, are notoriously difficult to apply with complete confidence that they will be certain of identifying all cracks above the assumed sensitivity level. Added to this are the uncertainties arising from testing a huge number of axles, in order to identify the very small sample of the population that may be cracked. Detection sensitivities are usually based on what size of crack a particular method might detect in a test-piece known to contain a crack. This is a very different situation from detecting which particular axle out of say 10 000 might be cracked. There is a suspicion that crack detection of axles is inefficient and if better understanding of the nature of what must be an 'extreme-value' fatigue event could be utilised, great savings on inspection may be possible.

1.4.4 Gearboxes, drive shafts, brakes, springs and suspension components

As we move further away from the wheel–rail interface, a whole range of components suffer from potential fatigue problems, and while load inputs arising from dynamic running loads are significant, their effect becomes more attenuated with distance from the origin at the interface. Cases of failures in all the components mentioned in this section heading have been reported in specific types of vehicle, but none could be said to be generic. In the past brakes generally operated by shoes acting directly on the tread of the wheel. There were counteracting effects: the wear produced by the action of the brake 'dressed' the wheel and rubbed out incipient fatigue damage. On the other hand, excessive braking tends to induce thermal damage at the wheel's running surface. Although brakes of this type still operate on older vehicles and on most freight vehicles, newer designs incorporate disc brakes, on which thermal crazing leads to spalling or fracture of the disc.

Brake pipes and connections are often made from rubber and rubber compound materials, the fatigue evaluation of which formed the basis of a study by Hansaka *et al.* (1999). Rubber springs, in blocks or formed into air bags, are frequently used in modern suspensions (Luo *et al.*, 2003), where they are subjected to fatigue deterioration. Most applications involving elastomers and other non-metallics require specific experimental testing programmes as the mechanical properties of such materials are generally not so well defined as those of metals and are sensitive to environmental deterioration and loading frequency effects.

1.4.5 Fatigue problems below the rail

The passage of a train over the rail and its supporting structure leads to potential problems in the rail fastening, the rail supports and the foundation of the track.

The modern method of fastening rails to sleepers is by metal clips, which in certain circumstances have failed. Sleepers, made from a wide variety of materials – wood, concrete, steel or composites – seem remarkably free from fatigue problems. However, ballast, the principal material used to make the track foundation, does suffer from continuous deterioration, which in the broadest sense may be classified as fatigue. Ballast is nothing more than a compacted pile of stones through which loads are transmitted by the contacting vertices. Wear at these points of contact causes settlement of the track and is the principal reason for the extensive and expensive maintenance needed to preserve the geometry of the top surface of the rail. It is not therefore surprising that continuous slab track, more expensive than ballast to install but much cheaper in maintenance costs, has been used on many modern railways.

This area has been the subject of a recent review (Dahlberg, 2001). Some empirical models of settlement of ballast were discussed. If settlement is expressed as a function of loading, either number of wheel passes or tonnage, a common feature is an initial rapid settlement blending exponentially into a longer-term and much slower constant settlement rate.

1.5 Fatigue and vehicles

1.5.1 Body shells

Historically carriages have been mounted on a heavy steel underframe with a relatively flimsy superstructure made from wood. There are numerous examples of the disastrous consequences of telescoping or overriding of such vehicles when the heavy underframe has mounted over the frame of an adjacent vehicle and penetrated the wooden passenger compartment. The change-over to all-steel bodies has led to a monocoque type of construction in which the whole tube of the carriage contributes to the structural strength. Although steels, including stainless, have been used in many designs, aluminium has been introduced relatively recently. The key advantage of the use of aluminium was the availability of long extrusions extending over the whole length of the vehicle, thus serving to simplify the construction. Many detailed fatigue problems at points of stress concentration have been corrected during the service of such vehicles, usually without great difficulties other than expense. Some concerns have been expressed about the low-energy unzipping of long welds in aluminium in the region of the heat-affected zone of the weld. Potentially, this could be dangerous in crash situations where the modern trend is to design structures to deform and crumple by absorbing energy, in a manner that is familiar in automobile structures (Smith, 1995).

Recent experience in the UK includes the withdrawal of a particular class of electric train after cracks were found in the underframe of some vehicles. A press announcement of this event included the remarkable statement that

> because they (the cracks) are in aluminium, in which cracks are not unusual, the cracks cannot be welded and engineers from the manufacturer are looking at a long-term solution which may mean replacing the sub-frame.

A more serious type of problem occurs in high-speed trains which must be airtight in order to protect passengers from pressure pulses during, for example, passage through tunnels or the passing of trains in the opposite direction. Small fatigue cracks growing from rivet holes of spot welds determine the useful life of such vehicles, which is considerably less than the traditional 40 years.

1.5.2 Engines, motors and couplings

It is obvious that the engines and motors used in railway applications are subject to the same generic fatigue problems as their static counterparts. Generally, electric motors are well behaved and, despite their increasingly small size allied to greater power density, offer very little trouble. Diesel engines, on the other hand, are worked hard in the railway environment, where the frequent discontinuous service contrasts with the steady operation in marine applications. Thermal cracking and associated fatigue problems are therefore relatively common. For all types of motors, problems arise with mounting bolts and brackets, problems which on investigation are generally straightforward to solve.

Couplings between motors and the drive train are exposed to high dynamic loads, and fatigue problems are common. Static couplings between vehicles see impact loads and variable loads during their duty cycle. Cracking is common, complete failure less so, but again the solutions are relatively straightforward.

1.5.3 Internal components and fittings

All the equipment contained within a vehicle is subjected to fatigue design considerations. This includes even the apparently trivial details such as broken handles, cracked plastic tables, broken hinges, toilet seats and the like. Although plastic components have played a valuable part in reducing the weight of fittings, detail design has often been inadequate to withstand the rigours of long-term use. Exposure to sunlight has led to loss of colour and sometimes strength. As is the case with automobiles, what was previously a basically mechanical product has been transformed to a complex mechatronics assembly. Information systems, air-conditioning units, diagnostic instrumentation and the like have been added to vehicles, sometimes cancelling out the efforts of structural designers to reduce mass. All the electronic equipment is subjected to the harsh mechanical environment of the vibrations and shocks generated by the motion of the train. In conjoint action with the thermal cycles associated with electronic equipment, the joints of circuits and components in such equipment are subjected to arduous conditions not experienced by static equipment. Most failures of electronic equipment stem from fatigue failures of internal circuit joints. The current requirements to remove lead from solder alloys add to the pressures caused by continuing miniaturisation and increased performance demands. The problems of structural integrity and reliability of electronic equipment are thus substantial and increasing (Plumbridge and Kariya, 2004). The scale of the problem can be judged from the estimated 10^{13} soldered electronic interconnections manufactured annually, about 1600 for every person on earth!

1.6 Fatigue in the infrastructure

Some aspects of the infrastructure, rails and track, with problems arising from the wheel–rail interface, have been discussed previously. This section turns to other aspects of the infrastructure.

1.6.1 Bridges

Bridges, in large numbers, form essential links in the permanent way of our railways. Their development encompasses the various building materials of the ages: wood, stone, brick, cast iron, wrought iron and steel. Their occasional failures have often been the stimulus to research into the proper use of materials, design and maintenance methods. Perhaps the most famous bridge failure in railway history is that of the Tay Bridge in 1879. It is unique among British railway accidents to this day in that there were no survivors. Recent re-examination of evidence suggests that fatigue may have played a key role (Lewis and Reynolds, 2002; Lewis, 2004). Even earlier, the collapse of the Dee Bridge at Chester in 1849, Fig. 1.6 (Lewis and Gagg, 2004; Lewis, 2007), led to a Royal Commission, and, *inter alia*, the first fatigue tests on large-scale railway bridge girders (Anon., 1849).

The maintenance of bridges, some well over 120 years old, is a major concern and cost for railway administrations. Corrosion is ever present on metal bridges, even if painting is continuous, as exemplified by the Forth Bridge in Scotland. Fatigue acts in an insidious manner, often conjointly with corrosion, sometimes separately. Masonry and brickwork are attacked by water and by colonising plants. Fatigue loads often need experimental measurement (dynamic stresses again being hard to calculate) and the guarantee of extended life requires the full-scale testing of components. The whole topic is extensive: the reader is referred to a recent review (Smith, 2004).

1.6.2 Signals and electrical supply components

Because of the long life required from railway infrastructure, traditional semaphore signals, operated by pulling a long steel wire, still exist in many parts of the railway world. Corrosion and wear are potential failure mechanisms, but examples are rare. Coloured lights are the most common modern signalling

1.6 The failure of the Dee Bridge in 1849 led to a Royal Commission on the use of iron in railway structures and almost led to the disgrace of the famous railway engineer Robert Stephenson.

method, now being coupled to increasingly complex electronic switching systems. The most modern high-speed trains are computer controlled with entirely cab-based displays. A detailed discussion of the failure mechanism of these vital systems is out of place here. It is, however, worth repeating that many failures of electronic equipment are caused by the failure of joints subjected to cyclic thermal stresses.

Modern electrified systems generally rely on an overhead wire for current supply. The dynamic behaviour of the wire requires delicate control to ensure good contact conditions between the wire and the electric current collecting device on the train. High power densities exist at the contacts, which can restrict the choice of materials used for both the wire and collector. The major deterioration mechanism is wear, although fatigue may play a part. Wires and their supporting structures, which are significantly fatigued, must be carefully monitored to prevent collapse of the supply system.

1.7 Concluding remarks: the future

It is tempting to say that, in general and despite recent high-profile accidents, we have sufficient fundamental knowledge of fatigue to operate railways safely. This comes, of course, at the price of external vigilance. The cost of inspection and maintenance is extremely high and in common with other industries, ways are always being sought to reduce these costs. Railways are increasingly under pressure to improve their economics – many previously nationalised railways have been privatised with this aim in mind. New technologies are being introduced; improvements in track have been mentioned previously; improved and automated inspection techniques, although not discussed in this chapter, are being developed, and this area, applied to both infrastructure and vehicles, merits continuing efforts.

The development of high-speed railways has brought increasing pressure to reduce the mass of vehicles. The technologies of the aerospace industry are increasingly being adapted, in materials (aluminium and composites), manufacturing methods and inspection techniques. Because the unsprung mass plays such a major role in the generation of dynamic stresses, its reduction is vitally important. But the wheels, axles and bogies which make up this mass are vital to the integrity of the train and, as we have experienced, their failures can be catastrophic. The safety margins of these components are therefore becoming less than the traditionally generous ones typical of the railways of the past. Allied to the light weighting needed to reduce dynamic stresses, there exist societal demands for improved crashworthiness of railway vehicles. Compromises are needed to ensure crashworthiness is not gained with the penalty of increased mass.

There still remain surprising gaps in our knowledge of the actual service stresses experienced by wheels, axles, rails and other key railway components. Measurement technology and analysis methods have now advanced to a stage where the experiments needed to generate realistic data are relatively cheap and straightforward. It is desirable that programmes of work to establish service stresses, particularly dynamic stresses, are conducted in the near future.

Thus some key fatigue issues in structural integrity remain for the railways of the future, which are different from and more challenging than those of the historical railway network which in their time prompted research and investigation that have become the cornerstones of our current knowledge.

1.8 References

Anon. (1849), *Report of the Commissioners Appointed to Inquire into the Application of Iron to Railway Structures*. London, Her Majesty's Stationery Office.

Anon. (1920), Fatigue phenomena in metals, *Scientific American Monthly*, **1**(3), 221–226.

Anon. (2004), *The Future of Rail*, London, Department of Transport, Cm. 6233.

Beaumont W W (1879), Discussion to: The strength of wrought-iron railway axles (Andrews T), *Trans Soc Engs* (October), 143–178.

Dahlberg T (2001), Some railroad settlement models – a critical review, *Proc Inst. Mech Engrs, Part F*, **215**(4), 289–300.

Esslinger V, Kieselbach R, Koller R and Weisse B (2004), The railway accident of Eschede – technical background, *Engineering Failure Analysis*, **11**(4), 515–535.

Hansaka M, Ito M and Mifune N (1999), Investigation on aging of train rubber hose, *Quarterly Report of Railway Technical Research Institute (RTRI)*, **40**(2), 105–111.

(All the papers in this edition of the *Quarterly Report of RTRI* are concerned with railway fatigue problems. Likewise the edition **45**(2), 2003.)

Hill S J and Everitt D R (1988), The service operating environment – a vital input, in Marsh K J, *Full-scale Fatigue Testing of Components and Structures*, London, Butterworths, 278–293.

Jack I (2001), *The Crash That Stopped Britain*, London, Granta.

Lewis P R (2004), *Beautiful Railway Bridge of the Silvery Tay*, Stroud, Tempus.

Lewis P R (2007), *Disaster on the Dee*, Stroud, Tempus.

Lewis P R and Gagg C (2004), Aesthetics versus function: the fall of the Dee Bridge, 1847, *Interdisciplinary Science Reviews*, **29**(2), 177–191.

Lewis P R and Reynolds K (2002), Forensic engineering: a reappraisal of the Tay Bridge disaster, *Interdisciplinary Science Reviews*, **27**(4), 287–298.

Lundén R (ed.) (2003), Special issue on wheel–rail safety, *Fatigue and Fracture Eng Mater Struct*, **26**(10).

Luo R K, Cook P W, Wu W X and Mortel W J (2003), Fatigue design of rubber springs used in rail vehicle suspensions, *Proc Inst Mech Eng, Part F, J Rail and Rapid Transit*, **217**(F3), 237–240

Moreau R A and Peterson L (1955), Field testing of diesel locomotive axles, *Proc SESA*, **XIII**(2), 27–38.

Murray A (2001), *Off the Rails*, London, Verso.

Mutton P J and Alvarez E F (2004), Failure modes in aluminothermic rail welds under high axle load conditions, *Engineering Failure Analysis*, **11**(20), 151–166.

Plumbridge W J and Kariya Y (2004), Structural integrity in electronics, *Fatigue and Fracture Eng Mater Struct*, **27**(8), 723–734.

Schijve J (2001), *Fatigue of Materials and Structures*, Dordrecht, Boston, Kluwer Academic Publishers.

Schijve J (2003), Fatigue of structures and materials in the 20th century and the state of the art, *Int J Fatigue*, **25**(8), 679–702.

Smith R A (1990), The Versailles railway accident of 1842 and the beginnings of the metal fatigue problem, in *Proceedings Fatigue 90, Fourth Int Conf on Fatigue and Fatigue Thresholds*, Hawaii, eds Kitagawa H and Tanaka T, Materials and Component Publications, Birmingham, **4**, 2033–2041.

Smith R A (1995), Crashworthiness moves from art to science, *Railway Gazette International*, **151**(4), 227–230.

Smith R A (2001), Rolling contact fatigue: What remains to be done? in *Proceedings World Congress on Railway Research*, CD Rom, Köln, DB.

Smith R A (2004), Railway bridges: some historical failures and current problems, in *Progress in Structural Engineering, Mechanics and Computation*, ed. Zingoni A, Leiden, Balkema, Book of Abstracts and CD Rom.

Smith R A and Hillmansen S (2001), Monitoring fatigue in railway axles, in *Proc 13th Int Wheelsets Conf* (CD Rom), Rome.

Stanzel-Tschegg S (ed.) (2002), *Proc Int Conf on Fatigue in the Very High Cycle Regime*, Vienna, 2–4 July 2001, *Fatigue and Fracture Eng Materi Struct*, **25**(8–9), 725–896.

Suresh S (1998), *Fatigue of Materials*, 2nd edn, Cambridge, Cambridge University Press.

Wöhler A (1858–1871), *Z. Bauwesen*, **8** (1858) 641–652, **10** (1860) 583–616, **16** (1866) 67–84, **20** (1870) 73–106 (original reports in German); an account in English was published in *Engineering* **11** (1871) 17 March, 199–200, and subsequent issues.

Wolmar C (2001), *Broken Rails*, London, Aurum Press.

2

Fatigue in railway and tramway track

L. LESLEY, formerly of Liverpool John Moores University, UK

Abstract: This chapter reviews fatigue in railway and tramway track. After discussing the excitation mechanism in causing fatigue, it analyses railhead failures such as gauge corner cracking, corrugations, side wear and rolling contact fatigue. It then discusses rail failures such as star fractures, fishplate failure, weld and tension failures. The chapter also considers rail fixing, sleeper and ballast failures. The penultimate section reviews potential failures in buildings and earth structures such as embankments, cuttings and shelves. The final section looks at fatigue issues specific to tramways and light rail.

Key words: fatigue in rail track, tramways, excitation mechanism, railhead failure, embankments, cuttings

2.1 Introduction

This chapter comprises 10 sections, which discuss the mechanisms for fatigue and failure in railway and tramway structures under vehicle and wheel loadings. Section 2.2 reviews the development of railway infrastructure. Section 2.3 considers the excitation and propagation mechanisms and their effects in the railhead, where wheel pressures are highest, and through the track into the foundation formation, and then into the supporting structures and subsoil.

Section 2.4 examines railway and tramway track structures. Section 2.5 looks at railhead failures. These include gauge corner cracking, rolling contact fatigue and longitudinal corrugations, with respect to mitigation and avoidance measures, and other railhead fatigue problems including side wear and derailment. Section 2.6 discusses fatigue behaviour mechanisms in the whole rail, including star fractures at rail ends, weld fractures and fishplate failures. In Section 2.7 the effects of fatigue below the rail are considered in fixings and support structures, including slab track.

For conventional railways, Section 2.8 analyses subrail fatigue failures in sleepers and ballast. Section 2.9 looks at earth structures: embankments and cuttings. Section 2.10 reviews major structures like bridges, viaducts and tunnels.

In Section 2.11, tramways and light railways are considered as separate rail systems that exhibit fatigue problems not found on railways, an especially important subject with the growth of new light rail systems and their costs.

This chapter also reviews the ways to avoid fatigue problems at the design and construction stage, and the methods available during maintenance to mitigate or prevent failures from fatigue excitation forces.

2.2 Development of railway infrastructure

Railways took up and developed, from earlier roads and canals, the idea that engineering and ground works could make alignments easier or more economical. In the eighteenth century civil engineers like Telford, MacAdam and Metcalfe (Robbins, 1965) improved and built roads and canals. They introduced constant gradients to ease the burden on draught horses hauling wagons up hills on new roads. The need for civil infrastructure was even more important for canals where maximising the length at the same level (contour canal) saved water consumed by locks in changing levels, reduced transit times and increased barge traffic capacity.

Later, for the new nineteenth-century technology of railways, civil engineers like Stephenson, Brunel, Locke and Vignoles used constant gentle gradients to enable low-powered steam locomotives to haul trains over hilly and mountainous terrain. Some of the issues involved in designing these civil infrastructures were topographical, geological and hydrographical surveys. In the days before computers and photography, this involved visiting sites on foot or horseback, and making detailed notes and drawings in manuscript. Engineering data was obtained with chains for measurements and theodolites for heights. Hand calculations, aided by slide rules, occupied pages of civil engineers' notebooks and journals. In practice civil engineering software today produces only marginally better designs but much more quickly, enabling more alternatives to be considered and allowing more data to be analysed.

In most cases in the nineteenth century, the civil engineer who designed the railway was also the supervising engineer, who had to make judgements on less than perfect data, especially in relation to the siting and application of structural forces. For tracks at grade, on the normal ground level, the mass and dynamic loads of passing trains comprise the significant load into the ground. For most civil infrastructures, the mass of the structure (e.g. embankment, viaduct, etc.) is orders of magnitude greater than the mass of passing trains.

Designing civil infrastructure therefore requires a thorough knowledge and understanding of ground conditions and behaviour, since where train loads at grade can be accommodated by the ground, an embankment or viaduct may surcharge, or overload, the ground beyond its bearing capacity. Here techniques for distributing loads into the ground, e.g. by corbelled foundations (Fig. 2.1), or into stiffer strata by piling, can enable a better alignment to be achieved economically than at grade.

One common technique for achieving a constant gradient up hills, e.g. to a pass between valleys, is to cut shelves. Here the soil excavated from cutting into

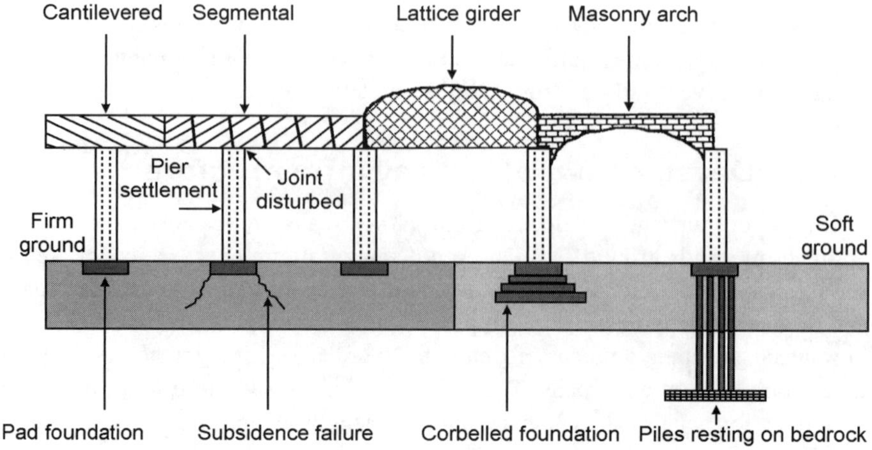

2.1 Types of bridge and viaduct design showing differing techniques for distributing loads into the ground.

the hillside is normally used to create an embankment on the downhill side, creating a shelf wide enough for the trackbed (Fig. 2.2). The slope of the cutting and embankment depends on the stability of the uphill ground to resist slips, and of the downhill side to accept the extra load of the embankment. Different ground conditions can be accommodated by changing the angle of the slope (batter) created by the cutting or embankment. Less steep slopes require larger volumes of spoil to be moved.

An example of an economical but sophisticated solution can be seen at the Horseshoe Curve in Pennsylvannia, completed in 1854 (Fig. 2.3). Here the

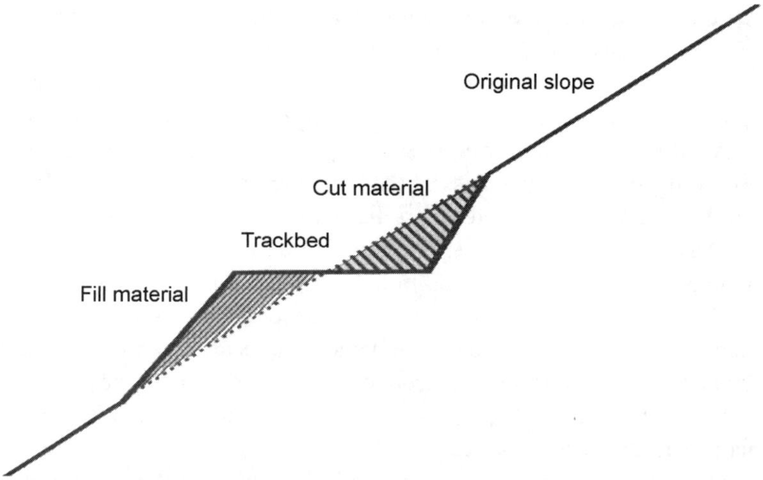

2.2 Design of shelf to accommodate railway track on a slope.

2.3 Design to accommodate a gradient: Horseshoe Curve, Pennsylvania, USA.

railway alignment climbs at a steady gradient of 2% to achieve a 300 m rise in 10 km. Midway up the gradient is a curve of 190 m radius around the head of the valley. Here the gradient is less severe, so that the rolling resistance of long freight trains (2000 m) is constant, whether the wagons are on the gradient or on curved sections. These gradient/curve combinations were calculated manually.

Failures of railway civil infrastructure are normally due to ground or structural problems, although harmonic loads from passing trains or peak wheel loading can accelerate or catalyse a failure. Such failures include cutting or embankment slips, tunnel roof falls, viaduct pier subsidence and arch spreading, and flooding.

2.3 The excitation mechanism

The motion of rail vehicles along tracks induces harmonic forces and resonant excitation. The principal mechanisms are vehicle speed, wheel eccentricity and railhead imperfections. Excitation frequencies induced in rail tracks are directly proportional to vehicle speed (v) (Fig. 2.4) (Lesley, 2000):

$$f = kv \qquad 2.1$$

where k = constant.

The excitation frequency (f) is determined by a number of factors. These include wheel diameter and degree of eccentricity (usually of the order of 200 μm). Excitation forces can be amplified by the resonant characteristics of vehicle suspensions, especially with worn dampers (Fig. 2.5). The magnitude of harmonic forces can be increased by wheel flats (of the order of 5 mm from diameter), and at constant velocity the interaction between sleeper spacing (hard spots) and vehicle suspension resonance, inducing secondary excitation.

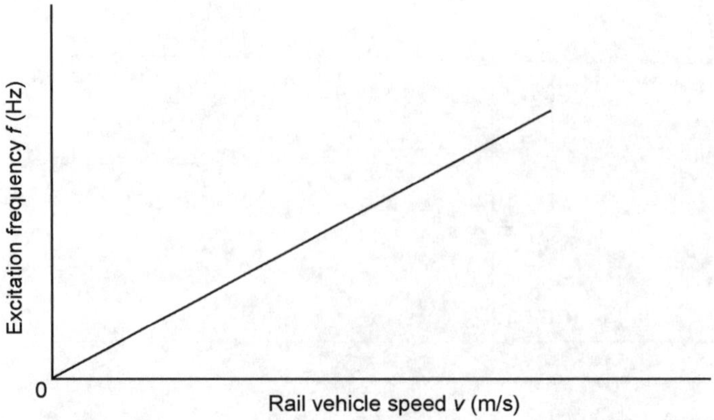

2.4 Rail vehicle speed and excitation frequency in rail track.

Finally, there is a second and interdependent excitation mechanism, due to lateral instability of wheel and fixed axle sets at speed (hunting). Hunting is the uncontrolled oscillation of wheel sets or bogies riding across and colliding sideways into the rails. Damping in the form of leaf springs or shock absorbers (vertical and yaw) can restrain the amplitude of hunting oscillations. The fundamental cause of hunting is the profile of wheel tyres and the restoring torsional axle forces to perturbations from a trajectory parallel to the rails. This is shown in Fig. 2.6. The choice of angle ϕ, the inward inclination of rails and the profile of wheel conicity determines both the lateral displacement and the frequency (F) of hunting across the track gauge

$$F = B(v) \tag{2.2}$$

where B is a tuned operator on v (train speed).

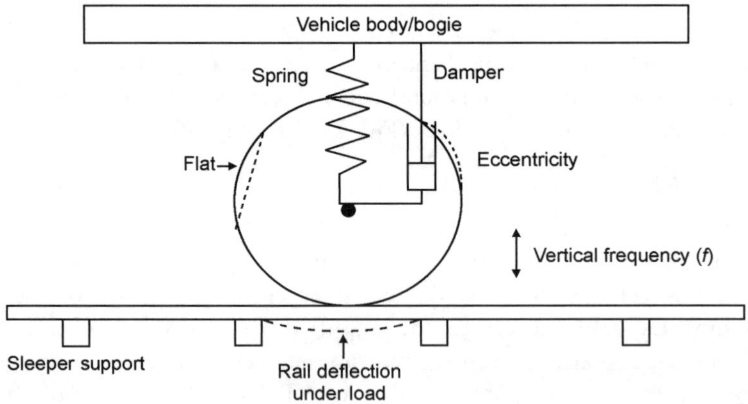

2.5 Impact of rail wheel eccentricity and suspension on excitation frequency in rail track.

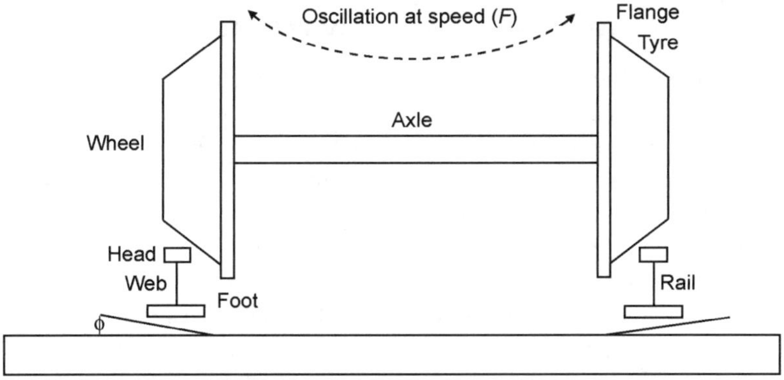

2.6 Rail wheel hunting.

Considerable research since the 1960s has produced a better understanding of the mechanism that initiates hunting. There are still factors that are difficult to model, including differential wheel and rail wear, the control of yaw in bogies, the torsional stiffness of axles, and the interaction between longitudinal and hunting excitation. Hunting is speed sensitive, with bogies being stable throughout the vehicle speed range, except at the resonant speed, where hunting occurs. (Fig. 2.7) For rail vehicles where wheels are not rigidly attached to a common axle, hunting does not occur. Research in India, Germany and the UK shows that independently rotating wheelsets do not create conditions of hunting (Lesley, 2000).

The impact of cyclic loading will result in fatigue failure depending on the level of stress and number of cycles, since failure is dependent on both (Hecht, 1994).

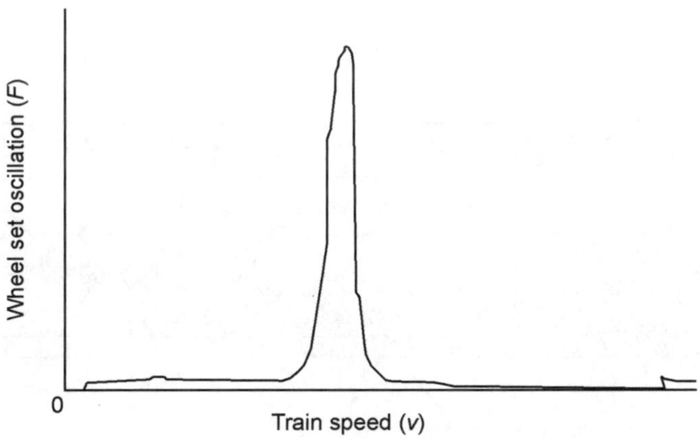

2.7 Relationship of rail wheel hunting resonance frequency to train speed.

2.4 Railway and tramway tracks and structures

There are significant differences between the structure and behaviour of railway and tramway tracks to warrant separate consideration (see also Section 2.11). One major difference is that railway tracks are used exclusively by rail vehicles, whereas tramways are often shared with road vehicles. These functional differences define the structural form of the tracks (Fig. 2.8). In terms of design, the rails for railway tracks are normally laid with a 1 in 20 inward lean ($\phi \sim 3^{\circ}$), while tram rails stand vertical.

Railway tracks (rails and sleepers) are laid on an elastic base of crushed rock (ballast), which supports and drains (Timoshenko and Langer, 1932). The design of the track aims to reduce the high wheel stresses on the rail, into lower pressures into the ground. While this is a low first-cost option in universal use, it does require constant and expensive maintenance if an acceptable and safe ride

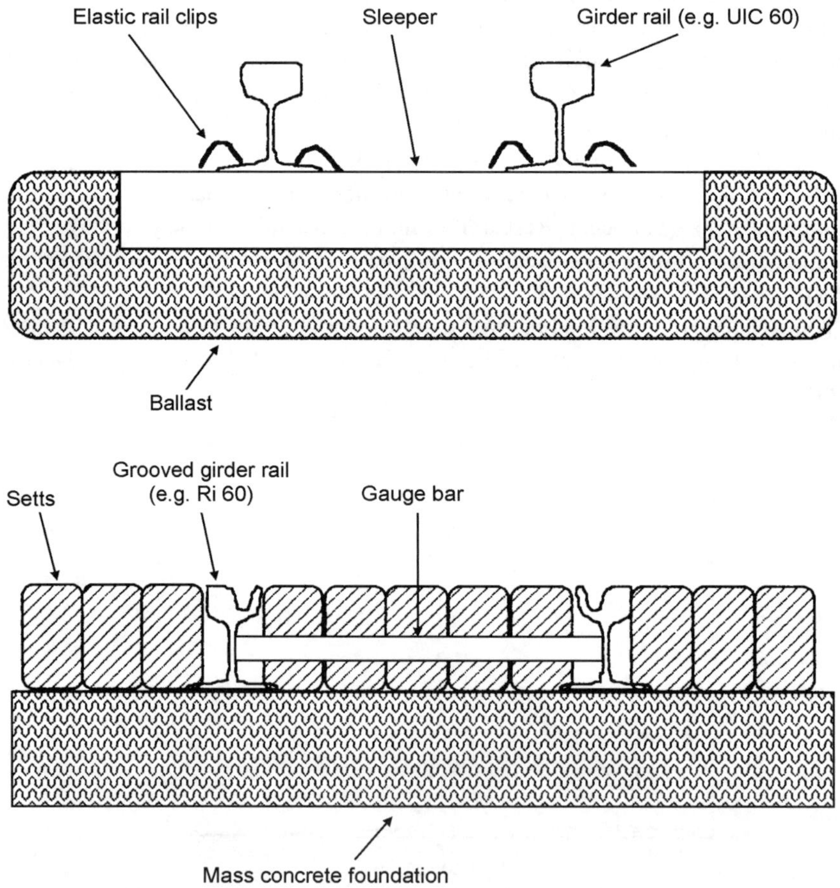

2.8 Cross-section of basic railway (top) and tramway (bottom) track design.

is to be provided. The maintenance costs are a function of maximum train speed and axle loading, being proportional to the square of both. The main maintenance is the repacking of the ballast (tamping) and ballast cleaning for drainage and elasticity. Work on the optimum depth of ballast shows that a ballast depth of more than 500 mm does not improve ride quality or reduce maintenance costs. Normally a ballast depth of about 300 mm is used.

Considerable development work has been undertaken with the use of slab tracks, which can be economically laid for new lines but have a different behaviour from ballasted sleepers. In particular, slabs are rigid structures with the rails continuously supported. Slab tracks have found their widest use for structures like tunnels and viaducts, where the slab can be part of the load-bearing structure. Replacing existing railway lines with slabs, as presently conceived, is often impractical due to the long period required for concrete slabs to cure. The use of pre-cast track slabs has not been widely considered, because of the cost and difficulty of transporting to site.

Tram rails have historically been laid directly on mass concrete foundation slabs (usually about 500 mm deep and 2500 mm wide per track), with the highway pavement made up to the railhead, including a groove to accommodate tramcar wheel flanges (Fig. 2.9). Tramway tracks are excited by both rail and road vehicles. The latter has been the cause of some notable track fatigue failures (Lesley and Al Nageim, 1996). These failures are often due to the interaction between rigid tram tracks in flexible highway pavements, creating fatigue between the two from the excitation of road vehicle wheels.

A further difference is that rail tracks are designed for high maximum train axle loads (e.g. UK 25 tonnes, EU 22.5 tonnes, US 32 tons, etc.). Whereas tramcars tend to have lower maximum axle loads (about 10 tonnes), tram tracks are impacted by road vehicles with up to 18 tonne axles. There are examples of mainline rail vehicles (including freight) using tramway or light rail tracks for collection and delivery, and an increasing number of tram systems share heavy rail tracks (e.g. Karlsruhe, Saarbrücken, San Diego, Sunderland, etc.) (Lesley, 1996; Matsuura, 1992).

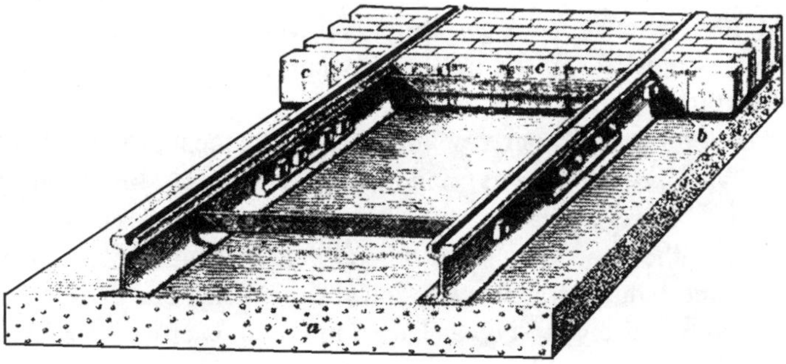

2.9 Nineteenth-century tramway track design.

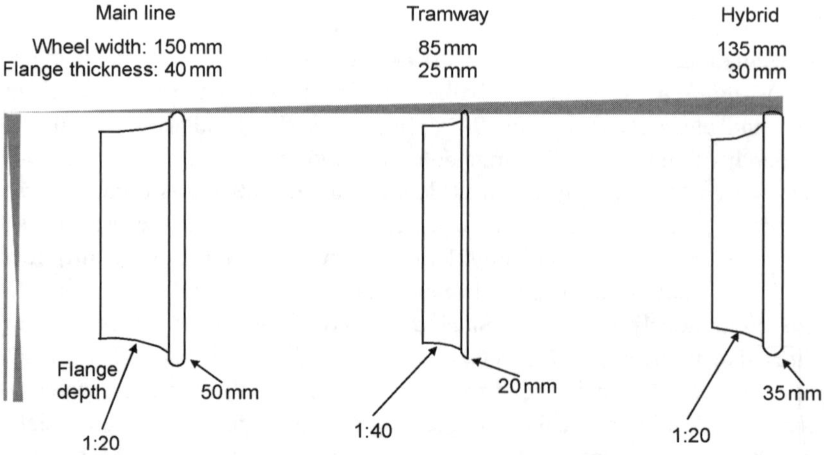

2.10 Main classes of rail and tram wheel profile.

Track excitation mechanisms are further complicated by the large variety of rail and wheel profiles. The latter can be subdivided into three main classes: mainline, tramway and hybrid (Fig. 2.10).

2.5 Rail head failures

The excitation mechanism and wheel dynamics already discussed lead to four main fatigue-induced railhead failures: gauge corner cracking, corrugations, side wear and rolling contact fatigue, depending on levels of stress (ORE, 1966; UIC, 1979).

2.5.1 Gauge corner cracking

This is a speed-induced shear fatigue force acting laterally across the railhead, leading to micro-cracking through the railhead (Profillidis, 2000). *In extremis*, cracks propagate right through the rail which fails, as occurred at Hatfield, UK, in October 2000 when a passenger train derailed at a fatigue-fractured rail. This phenomenon is most severe on tight radius curves (<1000 m) and where trains run at over 100 km/h.

The cause of gauge corner cracking is a consequence of the behaviour of fixed wheel/axle sets (in bogies) negotiating curves. At slow train speeds the phenomenon of wheel squeal is well known. This is a non-damaging force. At higher speeds, the wheels and rails continue to squeal but at ultrasonic frequencies, since squeal frequency is speed dependent. The cause of the squeal was postulated by Lesley in 2005 as lateral micro-slip across the railhead, as the wheelsets and tyres try to follow the track curvature (Fig. 2.11). Recent research in Italy has confirmed this hypothesis (Belforte *et al.*, 2006).

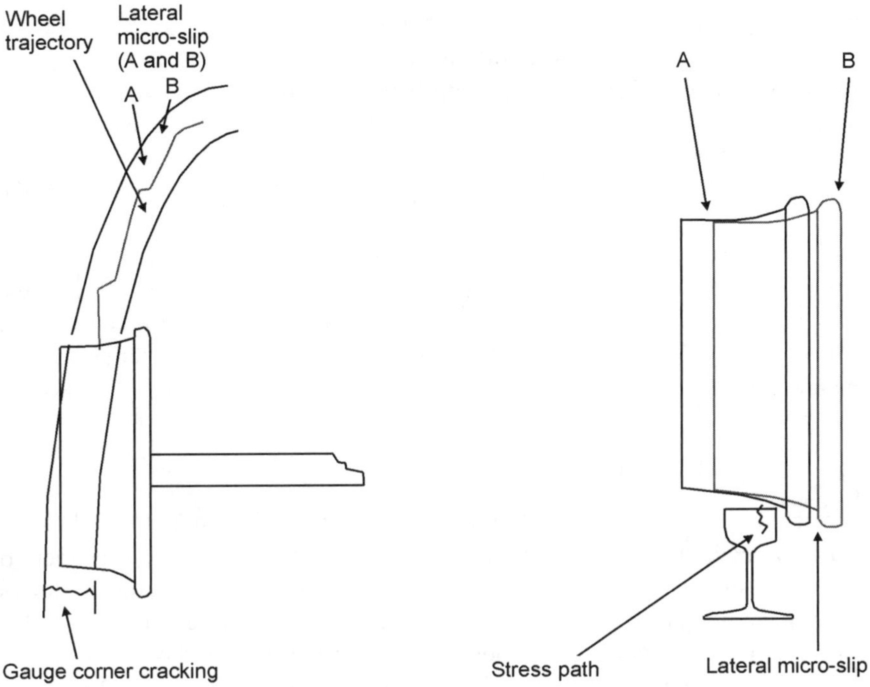

2.11 Rail wheel squeal and corner cracking.

On early nineteenth century railways (and most tramways), wheel flanges steered wheels and bogies around curves. Today railways use railhead and tyre forces for steering. With conical tyres and wheels fixed rigidly to axles, there is a restoring force when one wheel tries to rotate at a different speed from the other on the same axle on curved track. As with all such restoring forces, there can be an over-reaction which causes the wheels to slip sides periodically across the railhead. This micro-slip across the railhead induces shear forces and microscopic cracking in the running surface, from the highly harmonic lateral forces.

Upadhyay (2005) argues that the now universally adopted 'worn wheel' profile has increased the incidence of gauge corner cracking because of the prevalence of two-point wheel–rail contact on partly worn rails, and the compound stresses generated. His evidence suggests that the older, less coned wheel profiles were less likely to create the conditions of gauge corner cracking, since they are more likely to run with a single point of contact. The downside, however, is that the older profile is more prone to hunting.

Curved railway tracks are of course normally super-elevated, with the outside rail higher than the inside, providing a degree of cant (in the UK, up to 180 mm on 1435 mm track gauge). This, however, only gives a balance between centrifugal force from train motion against gravity at the design balancing speed. At other speeds, trains have a tendency to slip (outwards at higher speeds, inwards at lower speeds, especially for freight trains). So while high speed passenger trains have often been seen as the main cause of gauge corner cracking, all trains, even slower freight trains, contribute. Only by guaranteeing that trains always transit curves at precisely the balancing speed for the cant can the mechanism that causes gauge corner cracking be minimised.

As with other forms of fatigue cracking, the main treatment worldwide is the regular grinding of the railhead to remove embryonic cracks. This of course prematurely wears the rails, which require more frequent replacement than on straight tracks. It was the neglect of this regular maintenance that allowed the railhead cracking to propagate and cause a fatal failure at Hatfield in 2000. Kapoor (2002) sets out some strategies to address the problems of fatigue during maintenance.

2.5.2 Corrugations

Corrugations are longitudinal regular hardening of the railhead, with short (~0.5 m), medium (~2.0 m) and long (~6.0 m) wavelengths. These wavelengths arise from different excitation mechanisms. Short-wave corrugations are often found in station areas due to irregular braking or jerky acceleration (Fig. 2.12) Medium- and long-wave corrugations result from wheel set or bogie hunting. In all cases the fatigue effects in the railhead are the same. The peaks of corrugation coincide with hardened strips across the railhead, normal to the direction of motion (Lesley, 1996).

There are two results of corrugations. First, they create noise and noticeably uncomfortable riding, especially when train speed resonates the train/bogie structure. Second, without remedial measures of grinding (up to 2 mm depth) to remove the hardened areas, the hard zones propagate right through the railhead

2.12 Corrugations in rail track.

and become permanent, and then require complete rail replacement. While this is not a safety matter *per se*, it is a fatigue failure mode with cost and comfort issues.

2.5.3 Side wear

Straight tracks, or curves with large radii (over 500 m), and well-maintained rolling stock result in minimal side wear of rails. The wear of the railhead will dictate rail replacement periods. The fact that rarely will all rail vehicles be maintained in perfect condition means that straight tracks can suffer premature side wear due to bogie hunting and four-wheel vehicles crabbing, with wheels not parallel but angled across the rails.

Both these mechanisms cause shearing wear of the rail side, as well as impact damage (Fig. 2.13). If this is coupled to railhead corrugations, the rail side damage will also be harmonic and lead to hardening, compounding the damage to the rails; this is another example of a fatigue failure problem.

The importance of good rail vehicle maintenance can be illustrated by the wheel/rail behaviour work undertaken by British Railways Research in the 1960s, which led to the Advanced Passenger Train Programme. Fundamental to this was the identification of a wheel profile that produced optimal performance in terms of wheel steering. Indeed, this work showed that the tyre profile is the primary steering mechanism. Rail wheels can operate safely without flanges, given the correct 'worn' tyre profile. This means that in normal operation there would be no contact between wheel flanges and rail sides. Indeed, a rail vehicle project with which the author is associated, stable bogie and wheel behaviour, led in tests to wheel flanges rusting due to lack of contact with rail sides, even on curves under 50 m radius.

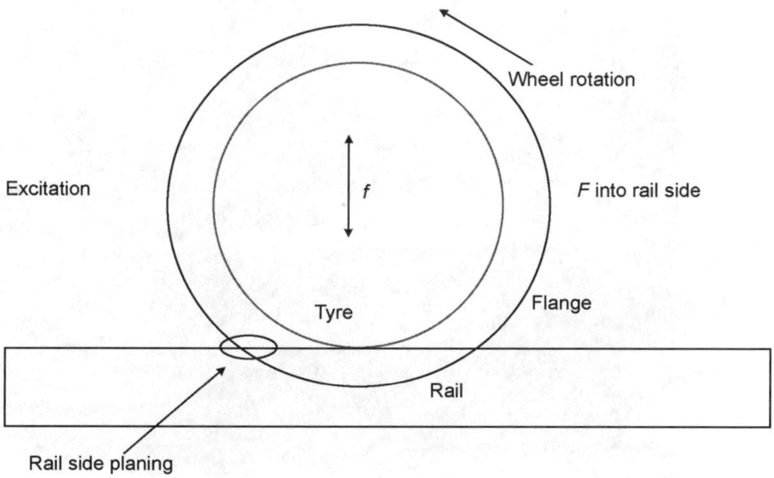

2.13 Mechanism of shearing wear of rail sides.

On smaller radius (<150 m) curves the tyre steering mechanism breaks down. Where hunting is also prevalent, wheel flanges will contact the rail side. This contact is both damaging and complex. The wheel flanges hit and grind the rail side harmonically, both laterally and vertically. Because of the shape of the wheel flange, there is also a slicing action, which planes off the rail side, inducing a wave formation in the side of the rail (Fig. 2.14). Because wheel flanges are not vertical, the side of the railhead takes an off vertical profile, which *in extremis* enables flanges to climb onto the railhead, causing the other wheel of the axle to fall off its rail and leading quickly to train derailment.

2.14 Wheel slicing action inducing a wave formation in the side of a rail.

2.15 Non-vertical side wear in a rail (photo: RAIB 2006).

Figure 2.15 shows an example in which wheel flanges have ground the rail side into a non-vertical profile.

Preventative measures include rail side lubrication, at places where flange grinding occurs, and wheel flange lubrication initiated by bogie rotation. While grease and wax are the most common lubricants, water will also perform this function adequately, without the problem of grease getting on the railhead. Greasy rails cause slip during acceleration or slide on braking, both of which also cause railhead damage, from burning (Fig. 2.16), where friction between wheel and rail, e.g. from wheel slipping, overheats the railhead and damages the metallurgical structure of the rail, which can then fail.

Once side wear has been initiated, side grinding to return the head to the correct profile is the first mitigating option. Premature rail replacement is the last resort. Side grinding (by wheels or maintenance) also creates a secondary problem of maintaining track gauge to safe tolerances. On straight tracks switching rails from one side to the other can also be considered, to use undamaged rail sides. This does not prevent further rail side wear but it does extend the useful life of rails.

2.16 Railhead damage from burning.

2.5.4 Rolling contact fatigue

The final fatigue problem on the railhead is rolling contact fatigue (RCF). This is caused by normal wheel rolling along the railhead and is a relatively newly diagnosed phenomenon. Before 1990 RCF had not been recognised, partly because it was confused with other forms of railhead fatigue but also because changes in rail metallurgy and rail speeds have made it more important. An example of the initial crack and propagation is shown in Fig. 2.17.

Ringsberg and Josefson (2002) suggested a modelling approach to explaining RCF, while Burston *et al.* (2002) developed an application model for the life of rails to control RCF. Similarly, Burston (2005) later proposed a way to predict the onset of RCF, and therefore to initiate remedial measures. Dembosky and Timmis (2005) reviewed the state of knowledge about the

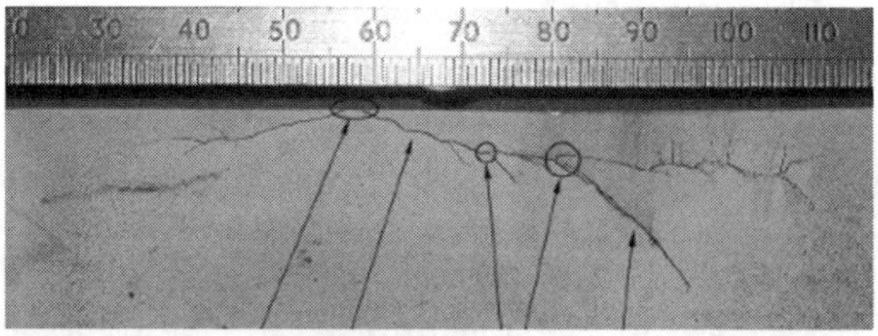

2.17 Crack propagation in a rail caused by rolling contact fatigue (RCF).

origins and progression of RCF. From a recent start, there is now a considerable body of both theoretical and practical knowledge. As with other railhead fatigue problems, if preventative measures fail, the only remedial measure is rail grinding to remove the onset of cracking, or, if left too late, the complete replacement of rails.

2.6 Rail failures

As well as fatigue cracking in the railhead already discussed, there are four other frequency-induced fatigue failure modes in rails; star fractures, fishplate failures, weld failures and tension failures (Fowler, 1976; Orringer *et al.*, 1984).

2.6.1 Star fractures

In Britain a significant proportion of tracks, especially on secondary routes, still use jointed track, rails bolted together with fishplates. To allow compliance for thermal expansion, the rail end bolt holes are oval, so that clamping bolts and fishplates can move relative to the rail (Fig. 2.18). Unlike in many other European railways, bolted rail joints in the UK are unsupported. As a train wheel passes from rail B to rail A, there is a vertical movement of A relative to B, fishplates and bolts. This harmonic vertical displacement causes micro-cracking to radiate from the bolt holes. Fortunately, it is a rare occurrence for these cracks to propagate right through the rail and cause a star fracture and rail failure (Figs 2.19 and 2.20).

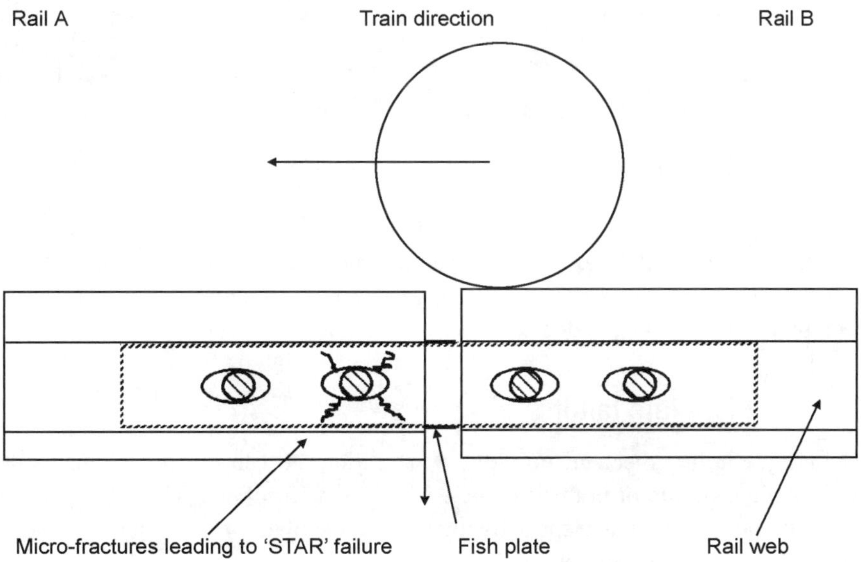

2.18 Star fracture due to micro-cracking radiating from bolt holes in a rail.

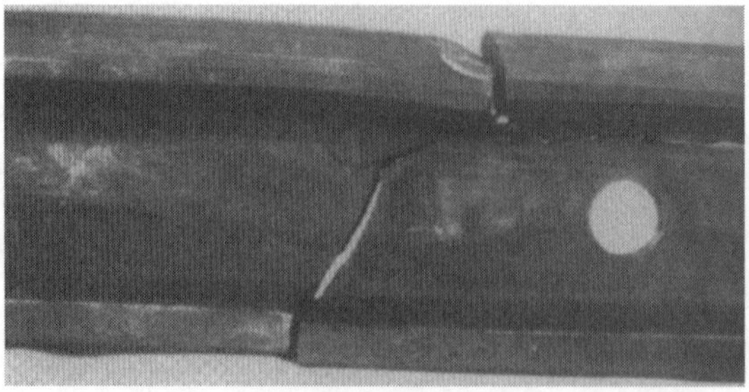

2.19 Example of a star fracture and rail failure.

2.20 Further example of a star fracture and rail failure.

The best remedial measure is for retaining bolts to be correctly fastened and greased, to minimise vertical play between rails, fishplates and bolts, while retaining longitudinal freedom for thermal expansion and contraction.

2.6.2 Fishplate failure

The fatigue failure mechanism outlined in Section 2.6.1 has an analogous mode in the failure of one or both fishplates, which secure rail ends (Fig. 2.21). Here the usual failure mode is a shear fracture of the fishplate, from periodic railhead shear pressure caused by passing wheels.

2.21 Example of fishplate failure in rail track.

2.6.3 Weld failures

Most main lines now have long welded rails. In spite of careful weld controls, the metallurgy of the weld will be slightly different from that of the adjoining rails. As these weld joints are normally unsupported, the passage of train wheels bends the weld joint like a beam, so that there is a vertical displacement, and tension in the bottom of the weld and compression forces in the top respectively. These, together with slight differences in railhead Brinnell hardness, stress the joint and can lead to two failure modes. The first is a fatigue tension failure at the bottom of the weld, which propagates upwards until there is complete separation. The second begins at the top due to shearing pressure and propagates downwards. These failures are harmonically induced, from the frequency of passing axles.

Fortunately, rail weld joint failures are rare (under 400 per annum in the UK and falling) and the most common cause is neither of the above but tension failures due to incorrect clipping of rails and over-contraction in cold weather. Regular ultrasonic testing can detect deteriorating rail welds and so allow repair or replacement before failure.

2.6.4 Tension failures

Rails are beams simply supported between sleepers and cantilevered at rail ends. At the point of train wheel contact on the rail, it is deflected with the upper surface in compression and the lower in tension. Away from the wheel contact, rails arch, the so-called 'bow wave', with the rail top in tension. Normally these stresses are well within the yield stress of the steel, with bow wave forces an order of magnitude smaller than the wheel deflection forces (Yasojima and Machii, 1965). An imperfection or premature rusting provides the seat for frequency-induced fatigue cracking to propagate upwards from the base, and

2.22 Premature rusting resulting in rail failure.

without replacement will lead to failure (Fig. 2.22). This failure mode occurred on 17 October 2003 at Hammersmith, UK, leading to a Piccadilly Line train derailing.

2.7 Rail fixing failures

2.7.1 Flat-bottomed rail

Flat-bottomed rails, fixed elastically to sleepers, allow the rails to rock sideways with the load of passing trains. This periodically compresses and relaxes the fasteners (e.g. Pandrol clips), which if not checked leads to the clip coming out of its seating, and releasing the rail. Fortunately a complete track failure rarely occurs. The loosening of clips is worse on tracks used by loaded four-wheel freight wagons, because of the large forces created laterally by wagon instability at speed, hunting across the track and pushing the rails out sideways.

2.7.2 Bullhead rail

The design of the cast iron chair, in which bullhead rails sit and are wedged with keys, prevents lateral rocking but not vertical displacement. Passing trains induce a frequency-dependent deflection. The keys work loose and drop out (Fig. 2.23). In the UK bullhead rails are keyed on the outside of the track, so that the loss of a series of wedges causes the rail to go out of gauge. In Ireland the keys are mostly on the inside of the track, simplifying inspection and making it less likely for tracks to go out of gauge if keys are lost.

2.7.3 Remedial measures

In both cases regular inspection is enough to prevent failures becoming catastrophic through fastenings becoming loose. Inspections should identify and

2.23 Loss of bullhead rail keys.

replace loosened clips and keys before rails can become dislodged, lose gauge and derail trains.

2.7.4 Switches and crossings

A couple of fatal derailments in the UK, at Potters Bar in 2002 and at Grayrigg in 2007, drew attention to the design and maintenance of switches (points or turnouts). In both cases the stretcher bars which held the moving point blades to gauge and alignment became loose, and in both cases also the nuts locking the bars to the blades were loose or missing. Clearly this is not a fatigue problem. However, there was also evidence that these stretcher bars were experiencing fatigue stresses from passing high-speed trains, which the derailment highlighted through breaking the stretcher bars where fatigue stress fractures had begun. This again draws attention to the need for proper and regular maintenance, since the blades would not have collapsed and the trains would not have been derailed had the stretcher bars remained firmly bolted to the moving rail blades.

2.8 Sleeper and ballast failures

Conventional ballasted railway track is designed to deflect elastically under train loadings, both from the total mass and from localised frequency-induced axle deflections (Savage and Amans, 1969). As in any elastic system, two mechanisms can cause fatigue failure (Zarembski, 1979). The first is the number and load of transiting axles, which flexes the sleepers and ballast. This creates fine particles by ballast grinding and these accumulate. This mechanism changes the track bed properties, which become less elastic and more rigid. The second comes into play when the elastic limit of the ballast drops below the maximum axle load. This creates another mode of catastrophic (plastic) failure of the ballast (Eisenmann, 1970).

These fatigue failure modes become accelerated when water is entrapped, creating lubrication that allows the aggregate and particles to flow and so reducing the effective depth of ballast, thus lowering further the elastic limit. In all cases, the timely cleaning or replacement of ballast after a specified train tonnage has been reached is enough to prevent plastic failure. Regular inspection to ensure that drainage is maintained is also important to prevent ballast flow.

For most railways, sleepers are made of wood, reinforced concrete or steel. The failure of sleepers is rarely a result of fatigue problems. The most common failure modes are due to material deterioration: wood rotting, concrete (reinforcement) spalling when internal steel reinforcement rods rust and delaminate the concrete, or steel sleepers rusting. These conditions also reduce the ability of sleepers to support the frequency-induced forces from passing trains into the ballast.

The design of the whole rail track is constantly under review, not just for the physical make-up of the track but also for its geometric properties. In one such approach Koc and Palikowska (2002) argue that train speeds can be increased on curves by adopting longer transitions, sometimes without the use of constant-radius sections. They show how this can be done within the existing alignment envelope using an iterative process, achieving significant transit speed increases.

2.9 Earth structures

2.9.1 Behaviour

Embankments and cuttings are designed and constructed to behave elastically under the cyclic loading of passing trains. Indeed, it is the passing train mass

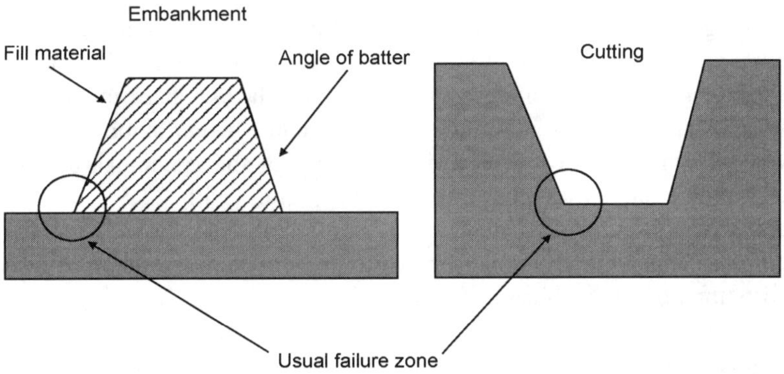

2.24 Cross-sections of typical railway embankment and cutting designs showing typical failure zones.

which is the main fatigue mechanism. Rarely, however, does the cyclic loading by trains actually cause fatigue failure, since there is a margin of over-design to accommodate unknowns in the subsoil on which embankments rest, or through which cuttings are made.

Typical design cross-sections are shown in Fig. 2.24. These also illustrate the most common failure but this is not usually caused or initiated by fatigue problems. Cuttings are dependent on the local ground conditions and the main design variable is the slope batter (slope angle), steeper in stiff soils and shallower in soft ground, requiring the removal of more spoil. For embankments the existing ground is a constraint in terms of its load-bearing capacity. The fill material can be spoil from nearby cuttings, imported quarried rock or synthetic materials like blocks of expanded polystyrene to reduce the pressure into the ground. In all cases the main failure mode is rotation at the base of the slope, which is not normally caused by cyclic or fatigue failure from train loads.

2.9.2 Design

Digital mapping and software (e.g. 'MOSS') can design vertical alignments that provide optimal profiles and minimise the need to cut and fill. It is easy to forget the immense skill of nineteenth century civil engineers who designed whole railways from their own field surveys and manual calculations. The principles, however, remain the same. The first stage in earthwork design is the definition of acceptable gradients and vertical curve radii. For railways with heavy haul (freight) trains, gradients under 2% are preferred. The reason for this is the adhesion limit of wheels on wet rails when restarting a train uphill or stopping a train downhill.

Train characteristics can have an important bearing on the need for and cost of earthworks. When French railways (SNCF) first designed their high-speed line between Paris and Lyon, it was envisaged that only one class of high-speed passenger train would operate, with low axle loads (17 tonnes) and distributed power bogies. A maximum gradient of 3% was adopted, which meant that less earthworks and no tunnels were needed. In places where tunnels would previously have been essential, hills were overcome by earthworks. In comparison, when German railways designed their first high-speed line, it was assumed that locomotive-hauled freight trains would operate with 22.5 tonne axles. This dictated less severe gradients (<2%). As a result, German high-speed lines have required considerable tunnelling, up to 25% of the alignment, at much higher costs than the TGV lines in France.

2.9.3 Embankments

Importing inorganic materials to raise the ground level to a new higher level results in the construction of an embankment. The mass of the fill material in an embankment surcharges the existing ground. This can be accommodated by using extra material to allow for ground settlement, or subsidence resulting from the surcharging, or by using less dense materials to reduce the surcharge to a pressure the ground can support with acceptable settlement.

The first stage in embankment building is to remove the topsoil and other organic material to reveal the subsoil, which should be less sensitive to environmental factors, including water levels. The cross-section of the embankment will depend on the nature of the fill material and the load-bearing strength of the subsoil. Raising the height of an embankment adds considerably to the volume of the fill material required (Fig. 2.25).

In constructing embankments, care must be taken to prevent voids being created in the structure, which will allow water to accumulate or trigger internal subsidence. The usual way to achieve this is by building the embankment with

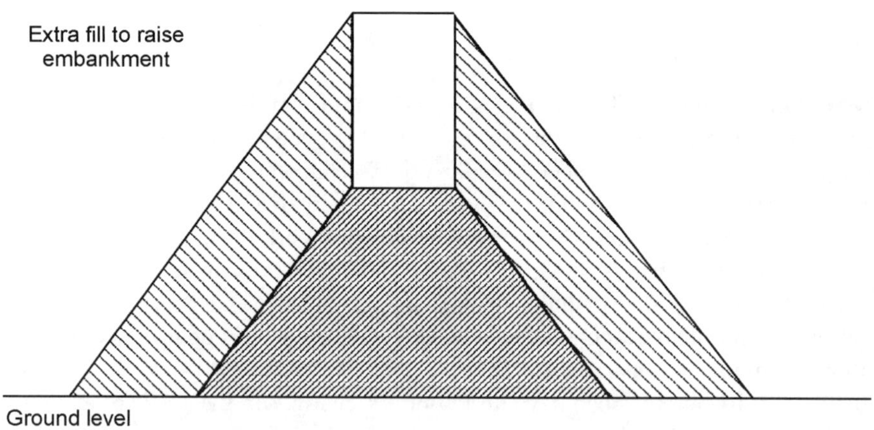

2.25 Raising the height of a railway embankment.

layers of fill material about 300 mm deep, which are then fully compacted to remove all voids, before the next layer is added. This was a lesson learnt from the first generation of railways with uncompacted embankments that frequently failed, even before the tracks were laid.

Where fill material available from local export (e.g. adjacent cutting or local quarry) is unsuitable, new material must be imported. In some circumstances a different approach can be considered using organic or synthetic materials as fill. Provided that this material is completely covered by an impervious layer (e.g. clay), such less dense material will remain inert and dry, and not rot or catch fire, since there is no air for combustion. Organic or synthetic materials (e.g. expanded polystyrene) will significantly reduce the ground surcharging while retaining the required load-bearing capacity. The load of a passing train on a narrow track is transmitted at a much lower pressure into the ground by the wide footprint of the embankment. Typically the pressure is 10 000 MPa at the railhead and 100 MPa at the base of the embankment.

Less dense fill materials can, however, be used only above the water table or possible flood levels, which would otherwise seriously damage or destroy the embankment due to the buoyancy of the fill. An early example of the use of buoyant material was the construction of the Liverpool to Manchester railway in 1828 across Chat Moss, a large area of deep waterlogged peat. George Stephenson, in failing to create a firm track by dumping rocks, used instead timber rafts which 'floated' on the peat, giving a wide footprint to distribute the load of trains into the weak peat (Morgan, 1971).

Embankments act like sponges on the local water table, and if not properly drained can cause falling rain to accumulate, with the potential for failure by washout, where the internal water pressure is greater than the strength of the embankment sides. This can be addressed in new constructions by the inclusion of adequate drainage out of and away from the embankment. For existing embankments remedial drainage may be required, by coring and using porous materials. Whilst embankments rarely fail as a result of fatigue, the cyclic loading from passing trains on a waterlogged embankment can be the trigger for a structural failure. Cannon and Henderson (2002) outline methods they have used to stabilise embankments showing signs of premature failure.

Embankments create physical barriers to other movements. These require bridges under the railway (under-bridges), culverts and other channels to let water, traffic, etc., pass across the line of the structure. When constructing new embankments, there are many techniques for including such breaks through earthworks. These can include masonry or concrete arches or decks, and pre-fabricated corrugated steel semi-cylinders (e.g. Armco) laid on strip foundations. In all cases, design care is needed in treating the transition from filled embankment to the under-bridge, where there is a significant change of structural stiffness and elasticity. This can be exacerbated by the harmonic action of passing trains, passing from 'soft' fill to 'hard' bridge and then back to 'soft' fill, creating differential compaction and changing the vertical alignment of the track. Transition concrete supports under the track bed are one way to mitigate this problem and to prevent differential settlement around the bridge which worsens the 'hard' spot.

For existing embankments, creating new ways under the track bed is more challenging, since the trains running over the embankment should not be disturbed. Fortunately there are a variety of 'no disruption' methods, based on pipe jacking. In these, prefabricated structures are pushed into the embankment on prepared foundations. Soil is then excavated from within the structure as it advances. Monitoring the top of the embankment is vital, to check for heave and more seriously collapse. Limits to the movement of the top of the embankment will be set to a few millimetres, after which trains will be slowed or stopped for investigation or remedial works. When such underpasses are perpendicular to the embankment, pipe jacking is easy, since the resistive forces are symmetrical about the front of the penetrating structure. Extra care is needed when such structures enter the embankment at an angle of less than 70º, when resistive forces are asymmetric. In all cases of pipe jacking through existing embankments, care must be taken to prevent the working front from collapsing and thus causing the embankment to differentially subside.

The most common embankment failure is rotation at the base (Fig. 2.24). The existing ground fails to support the load at the foot of the embankment. A section of embankment rotates into the ground, heaving the ground up and causing consequential collapse or subsidence of the side of the embankment. Repair of such collapses needs to consider the ground conditions that led to the failure.

Some of these failure modes, and resistance to fatigue failure, can be addressed at the design and construction phases and by the use of soil reinforcement. The simplest method for strengthening the ground includes mixing in proportions of lime as a stabilising material. For embankments, a geotextile can also be added to the construction layer of fill material. This stabilises the layers, increases the resistance to slope slip and minimises the mixing of particles between layers, thus maintaining the structural and elastic properties of the embankment. Finally, embankments can be pinned and piled into the subsoil.

The compound effect of an elastic track ballast on an elastic embankment needs to be understood when considering the likelihood of fatigue failure. The element with the lowest elastic limit will set the parameters for the ultimate strength of the embankment. The most likely system failure is in the track ballast, especially if measures have not been taken to prevent embankment materials contaminating the ballast. A contaminated ballast will cease to behave elastically, at which point it will fail by fatigue or plastic deformation and further stress the embankment, which in turn can fail. The use of a geotextile separation between ballast and embankment is one way to ensure that materials do not mix, and to encourage drainage to the sides of the embankment, reducing the risk of water percolating into the centre, where a catastrophic subsidence can begin.

2.9.4 Cuttings

Excavating a cutting through a hill to provide an acceptable railway alignment is usually a much cheaper option than tunnelling. Cuttings are also the second major earthwork in the civil infrastructure for railways. Some of the principles for embankments also apply to cuttings, except that the ground through which a cutting is made is already compacted. This, however, does not mean that problems due to geological strata and their inclination and faulting can be ignored. Relieving the overburden can lead to the subsoil relaxing and rising in height, with edge effects at the side of the cutting.

Groundwater and drainage are also likely to be significant issues at the design, construction and maintenance stage, and if ignored are a major cause of failure. One of the main problems with cuttings is drainage, in terms of flooding track and ballast. This reduces the life of the track. Groundwater is a cause of rotational slip at the bottom of cuttings and therefore a structural weakness. Such subsidence produces operational difficulties in terms of track deformation and train derailment. In some cuttings natural drainage will not be possible and excess water will have to be pumped out, especially in underground railways.

The main design criterion for a cutting is the stiffness of the subsoil. If this is uniform, then a uniform cutting batter (slope angle) can be achieved. If the subsoil is not uniform, by faulting or strata inclination, then the batter will be determined by the least stable stratum. Alternatively, civil engineering measures such as soil pinning, reinforcement or *in extremis* retaining walls will be needed (Fig. 2.26). These add to the cost and area of land needed for the cutting. Softer soils also create larger volumes to excavate and transport, even if only to an adjacent embankment. In deep cuttings, the ground exposed at the bottom, relieved of overburden, may itself relax elastically but not uniformly over some years. This relaxation needs to be accommodated in the design, both for alignment and also for cutting side details.

The track ballast in a cutting can more easily become contaminated by soil working off the cutting sides, especially if drainage is a problem, e.g. by springs appearing from a transected water table. This can be avoided by designing the bottom of the cutting with a gradient, or crest curve, so that surface water and groundwater will drain under gravity out of the cutting. The slope and size of drains will be determined by the volume of water expected. Where a cutting is continuously below ground level, e.g. for an urban metro, then drainage will normally be by mechanical pumping.

Unlike an embankment, where only the rain falling on it has to be drained, a cutting acts as a collector of rain and surface water from a large catchment area. Water ingress will be a major drainage problem, especially if the water table is

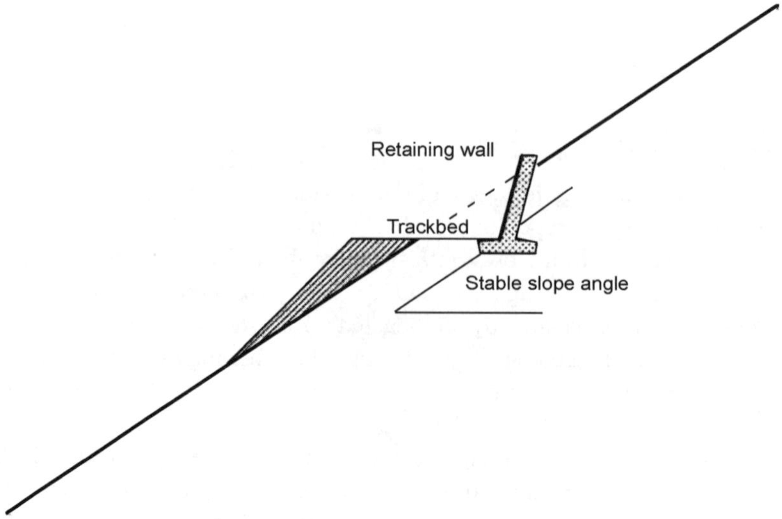

2.26 Retaining wall for a hillside cutting.

breached and springs appear. Track ballast that is continuously wet or under water will quickly deteriorate and fail.

Finally, cuttings, like embankments, disrupt surface routes and these have to be maintained by over-bridges and similar structures. Here the strength and stability of the cutting sides will determine the design of foundations and abutments to support the loading of traffic across the bridge. Building new cuttings across the line of existing routes means that their rights of way must be maintained. For some routes, e.g. canals and rivers, careful preparation will be required to maintain navigation. One way this can be achieved is by digging part of the cutting while constructing an aqueduct, and then diverting the river or canal. The old waterbed can then be excavated for the new cutting.

2.9.5 Shelves

Where a railway alignment has to climb up the side of a valley or hill, then a third earthwork, a shelf, is used (Fig. 2.26). Shelves contain elements of embankments and cuttings but often in more complex ground conditions. A cut into the side of a hill or valley may produce an angle of batter steeper than the natural stability of the existing slope. Measures to reinforce or stabilise the slope will be needed. These can include piling, pinning, gabions (surface interlocking blocks) or retaining walls. Some of these options were discussed earlier.

Material taken from the cut side will be used at the same location to provide the fill for the embankment, which is the other part of the shelf. Again soil stability and surcharging will dictate the angle of embankment batter and whether any measures of ground reinforcing or anchoring are needed, or even retaining walls.

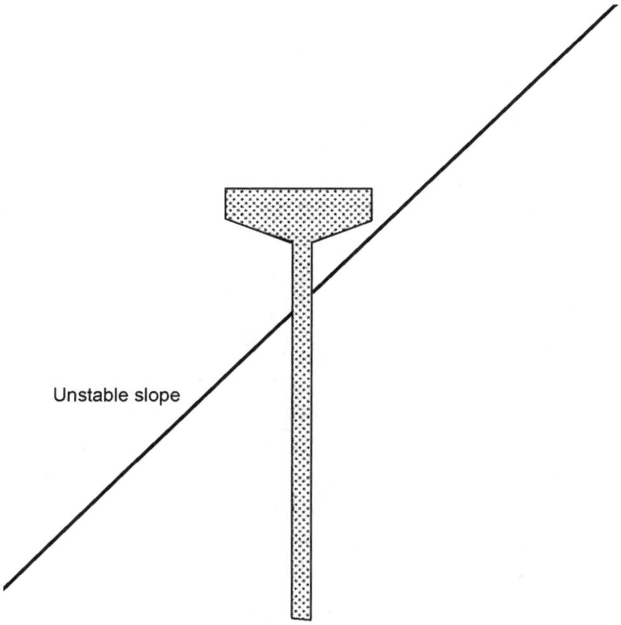

2.27 Piled viaduct to support rail track on an unstable slope.

In extremis, where the geology (e.g. strata inclination) makes a shelf impossible, viaducts or other structures that depend on piling and bedrock stability will be needed (Fig. 2.27).

The design and construction solution will depend on the ground conditions and water location. As with embankments and cuttings, fatigue is rarely the cause of shelves failing but can be a trigger for a failure due to more serious structural problems, e.g. soil stability, shelf design or construction detail, or drainage.

2.9.6 Stressing

Earth structures can be stressed for other reasons: high water table, retained water, washout, or subsoil subsidence. In these cases cyclic train loading can be enough to initiate a failure when the structure is less resilient. This fatigue or elastic limit stress is rarely the cause of failure but more often is a catalyst for failure from a more serious underlying structural problem.

For most earthworks, the mass of the structure itself will be at least an order of magnitude greater than that of the trains that pass over them. Earthworks will therefore stress the ground on which they stand, or through which they are cut. Passing trains may in practice cause no extra stressing of the ground. Good design and construction preparation are therefore important to prevent overstressing above the bearing capacity of the soil.

2.9.7 Failure

Regular inspections, especially in wet weather, to identify drainage problems, and prompt maintenance or repair are the most cost-effective ways of preventing catastrophic failures, which mean lines having to be shut so that the earth structures can be rebuilt. One harmonic excitation mechanism that can lead to earthwork failure but is outside the control of an operating railway is seismic activity. This can be due to natural causes like earth tremors, or man-made, e.g. from mine workings. This latter led to the diversion in 2003 of the East Coast Main Line near Edinburgh, where old mine working subsidence was creating problems in maintaining acceptable track alignment for high-speed trains.

An example of an earthquake failure occurred in northern Japan in October 2004, where the epicentre was immediately below the alignment of a high-speed railway line. The 6.4 magnitude earthquake failed to trigger train warning mechanisms but thanks to the reaction of the driver the train was stopped without any fatalities or serious injuries, although the front and rear carriages were derailed (Fig. 2.28) and the track was damaged. It is worth noting that the viaduct on which the tracks are laid was not damaged and, after track repairs, normal services were resumed within a day.

Another dramatic example of earthquake impacts on transport occurred in San Francisco in 1989, when the upper deck of the Cypress elevated freeway collapsed onto the lower road deck, killing many drivers and passengers. In contrast, the local metro (BART) trans-bay tunnel suffered no damage, no trains were derailed and a full metro service was resumed within two hours of the earthquake. BART was the only transport system in the area able to resume full operations immediately afterwards, without the need for repairs, because it had been designed to withstand earthquakes.

2.9.8 The future

New railways will be built. These will need earthworks and cuttings. For high-speed railways such earthworks are important in reducing journey times by the creation of direct routes with satisfactory vertical alignments and the maintenance of high operating speeds. Earthworks are an economical solution to building a railway through undulating surface topology.

Earthworks can be designed to give long and safe service, and for the most extreme vibration environments, including earthquakes. Train-induced vibration fatigue has rarely been the cause of an earthwork failure but can be a trigger, where there is already a structural weakness, either in the ground conditions or in the earthwork due to poor design or construction, or a lack of proper maintenance.

The train was travelling at 200km/h when the earthquake hit.

Bullet train derailed as quake sensors fail

TRACKSIDE EARTHQUAKE sensors failed to prevent derailment of a Japanese bullet train at the weekend as a Richter 6.8 tremor shook the line between Tokyo and Niigata.

But the train came to a halt without toppling over or causing injuries, despite the fact that it was travelling at 200km/h.

It is thought the earthquake epicentre was so close to ground level that on-board sensors were unable to react quickly enough to shut the train down.

Instead, the driver felt the force of the tremor and put on the brakes, bringing the derailed 10-car train to a halt.

The Toki 325 Shinkansen – or bullet train – was jolted from the track by the earthquake at 5.56pm local time on Saturday. The derailment was the first in the bullet train's 40 year history.

10m high viaduct section near Nagaoka on the Joetsu line.

Front and rear cars left the track while eight remained on the line. None of the 151 passengers aboard were injured, although some suffered shock.

Kazuhaku Aidi of operator the East Japan Railway Company (JR East) said it was fortunate that the train was travelling on a straight section of track.

He confirmed that the concrete slab track also had gulleys either side for snowmelt drainage. These are thought to have restrained movement of the derailed cars, preventing them from tipping over.

The train is thought to have travelled for several kilometres before stopping.

Toki 325 is one of the first generation Shinkansens and is heavier than later versions.

the derailment without serious damage.

Aidi said the line was protected by an Urgent Earthquake Detection & Alert System (UrEDAS). This uses seismometers at intervals to

earthquake and shut down the power lines.

But this may not have operated in time.

UrEDAS depends on picking up the "P-wave" of mainly vertical acceleration which arrives ahead of the horizontal component of a normal seismic event.

But Saturday's earthquake was particularly shallow. Professor Akira Wada, a structural engineering seismic specialist at the Tokyo Institute of Technology, said the epicentre was no more than 10km underground compared to a normal 40km or 50km. It was also almost directly below the train's location.

"This means the vertical acceleration component was particularly strong," he said. "And the two waves would not have been clearly separated."

He said reports indicated that the driver himself and braked the train.

JR East is working with the Japanese Land, Infrastructure & Transport Ministry and Tokyo's Railway Technical Research Institute to examine the train and track.

Adrian Greeman

2.28 Report of derailment in Japan due to earthquake damage.

2.10 Built structures

Railways require built structures, e.g. bridges, viaducts and tunnels, to carry tracks and give suitable operating alignments. These structures are also subject to cyclic loading from passing trains, high stresses from individual axles (especially wheel flats), thermal forces and other natural factors, including frost. Tunnels (Fig. 2.29) are also subject both to movements in the ground above the roof, which can lead to tunnel collapse, invert or foundation failure, and also to rotational failure at the foot of tunnel walls. The last group of railway structures are trackside facilities, e.g. stations, good yards, maintenance depots, etc., which are needed to load and unload trains, or to service rolling stock.

As with earth structures, fatigue is rarely a direct cause of structural failure. Monumental railway failures, like the Tay Bridge disaster in 1879 (Morgan, 1971), have all been due to overstressing. Vibration can be a catalyst for other

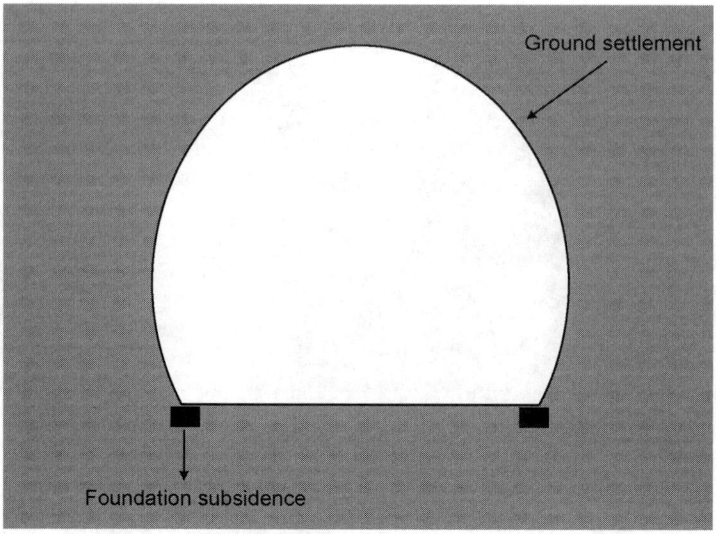

2.29 Typical failure zones in railway tunnels.

problems, principally water ingress and foundation settlement. Regular inspection and preventative maintenance, especially for proper drainage, are the best ways to ensure structural integrity, especially from frost damage. Like earthworks, seismic vibrations can cause fatigue failure, but this is very rare. Such structures, including viaducts, can be prone to the effects of ground movements. Hughes (2002) showed how the Yarm Viaduct was stabilised *in situ* after subsidence began to affect the piers.

The recent earthquake in Japan, see Section 2.9.7 above, caused no direct damage to railway structures. A viaduct was indirectly damaged by a derailing high-speed train. This damage was only superficial (*New Civil Engineer*, 2004). Designing the bridge for the location is particularly important for masonry bridges. However, with over 2000 years of experience in building masonry bridges, the railway age encouraged builders to stretch the technology to its limits, as with the spectacular Pangbourne Bridge built by Brunel over the River Thames in 1840. Brencich and Colla (2002) discuss these issues in terms of the mechanics of the bridges under load. Similarly, Harvey (2002) examined in detail the arches of masonry railway bridges to be able to predict remaining life under the cyclic loading of trains.

2.11 Tramways and light rail

Tramway tracks carry heavy road vehicles, imposing not only cyclic compressive loadings on tracks but also tension and shear forces. The destructive force of rubber tyres is amplified by water, which is compressed between joints in the tramway track and finds weaknesses leading to failure through 'pumping'. This forces water between tracks and pavement, and splits pavements and tracks off foundation slabs. The failure in 1993 of the tramway track in Moseley Street, Manchester, was caused by buses pumping surface water into the track structure. This has been a recurring failure problem. Poorly designed or installed tramway tracks can also fail when the groove check rail breaks off, because of either tram wheel flange action or heavy road vehicles like buses. This potentially creates the conditions where trams can derail in streets.

2.11.1 Switches

Locations where tracks divide or join are called switches, turnouts or points. In the facing direction the rail switch has a moving tongue to direct the flanges of tramcar wheels, and is the most vulnerable element of tram tracks. There are two reasons for this vulnerability. The first is that the radius of switches, being of the order of 20 m, means that even at slow approach speeds there are considerable lateral and vertical cyclic forces on the tongue. This is not a new problem and the 'tadpole' tongue developed in the early twentieth century (Dover, 1965) distributes forces over wider areas, reducing failures. The second is that all tramway switches depend on flange guidance, unlike railway points which depend on tyre guidance. Flange guidance grinds the side of the tongue and leads to rail climbing. The evidence for this is apparent at many tramway junctions, where wheel tracks can be seen in the highway pavement after trams have derailed. The southern arm of the Manchester Delta junction provides an example, but all tramways have switch derailments, since the profile of the wheel flange needs to be maintained, as these also are ground away on curves.

German tramways have developed an alternative to this, for operational reasons. This reduces switch wear and derailments by the use of pre-sorting at the approach to a tramway junction. Here the actual switch is set back some 50 m from the junction, allowing much larger radius (>100 m) curves, and a second interlaced track, with rails side by side. At the junction the switched tram only faces a plain track curve, with constant forces.

Road vehicles can be more damaging to tramway switches than trams themselves. The cyclic loading of switch blades by buses at the end of Moseley Street in Manchester is a cause of premature failure and tongue replacement (Howard, 2004). Maintenance can rarely ameliorate what could have been avoided at the design stage.

For trailing switches there are two designs. The first is a solid convergence of rails, but in such cases trams can only pass in the trailing direction. The other includes a permanently sprung tongue. This means that in the facing direction trams always take the same track. These are used at crossovers for reversing trams, without the need to reset switches or any other operating mechanism. This is another major operational difference between railways and tramways.

2.11.2 Crossings

These are the rail units that allow one rail or track to cross another. In a turnout from one track into two, there are two switch units and one crossing unit (Fig. 2.30). There are different design philosophies to this. In Europe until recently such turnouts had only one tongued switch, on the outer rail, with the inner switch being open. In North America the single moving switch tongue is usually on the inner rail, guiding and grinding the back of tramcar wheels. Practice in new UK light rail systems has been to mimic mainline railways with two tongued switches.

The life of crossings and failure rates depend on whether wheels are supported on their flanges over the crossing, or use the railway practice of tyre running. In the former case, the bottom of the rail groove is raised to catch the wheel flange and lift the tyre off the running railhead (so-called 'silent' crossing). This has the effect of reducing cyclic loading on the crossing nose and fatigue stress fracturing. The key to this is the profile of the approach ramp in the rail groove, so that cyclic loadings at the crossing are replaced by vertical loads on the approach. The torsional stress in tram axles can be reduced by lifting the groove and supporting the wheel flange on the other rail adjacent to the crossing. In any case, by distributing these loads away from the crossing, failures are reduced, as well as giving a quieter environment for residents and more comfortable transit for passengers.

Unfortunately many of the new light rail systems in the UK (e.g. in Nottingham) have adopted the railway practice of tyre running over crossings, which result in large cyclic loadings as the wheel falls into the crossing (frog) space and creates a noise nuisance to residents. Even with integral cast units this cyclic loading leads to fatigue fracturing. Fabricated units, except on lightly used lines, have a shorter life, as the cyclic loading weakens and fails welded or

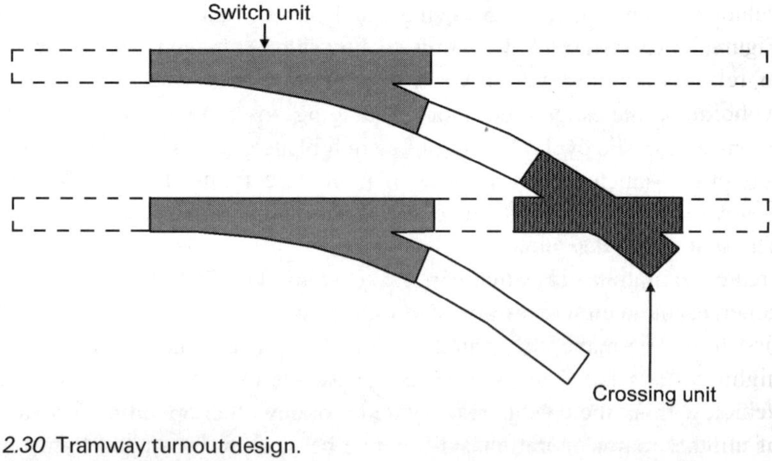

2.30 Tramway turnout design.

bolted joints. There is, however, one further crossing variant that avoids this but that is only possible when there is a main line and a lightly used branch line. Here the crossing on the main line has a continuous rail and groove. On the branch line, tram wheels climb over the main line. An example can be seen on the Blackpool tramway, at the Foxhall junction.

2.11.3 Embedment

Tramways traditionally have the highway pavement remade up to rail level using granite setts or similar blocks, or a flexible material like tarmac. Setts are prefabricated surfacing blocks, usually about 150 mm square and 100 mm deep, sealed with bitumen or tar, to provide a durable and waterproofed structure to support road vehicles using the same highway space. Setts were adopted as a simple method to access tracks when replacement was needed. At the time of construction in the nineteenth century, the operating life of rails was uncertain.

In the UK there is a legal anomaly that tramway operators are required to maintain the roadway between tram rails and 450 mm outside. Setts were retained as a durable road surface, when the rest of the highway was resurfaced with a smoother flexible material like tarmac. The interface between setts and tarmac is also a location for fatigue failure, when the tarmac edge ceases to behave elastically in contact with rigid setts, under loading from heavy road vehicles.

With the advent of tarmac and other flexible pavement materials, differential cyclic loading adjacent to rails is the principal cause of fatigue and pavement plastic failure. When this occurs there is an ingress of surface water, exposing the tramway track to structural failure. This has occurred in the Moseley Street, Manchester, tramway but only on the track shared with buses, not on the track used exclusively by trams.

One further fatigue problem for tramways arises when railway track designs are used, with elastic fastenings, embedded in a highway pavement. Here road vehicles wobble the rails, which then disturb and fail the pavement embedment. Figure 2.31 shows such a failure at a fabricated crossing. The need prematurely to relay curves on the Croydon Tramlink in 2006 was due to the tramcars wobbling the rails.

2.11.4 Tramway foundations

The first tramways were laid in unpaved roads. These tramways therefore had to create foundations on which to lay rails, in unpaved highways. The use of mass concrete foundation slabs was simple and economical. In the 170 years since the first tramways were laid, highway technology has developed rapidly, so that a highway pavement is itself a robust and durable structure for heavy road vehicles. In most cities, the space under streets has also been filled with a variety of utilities, ducts, pipes and cables.

2.31 Failure of pavement embedment adjoining tram track in Sacramento, USA.

The worldwide development of highway pavements has resulted in a substantial body of empirical knowledge on the performance of such pavements under defined vehicle cyclic loadings. Pavements can be designed to give lives of 30 or 40 years or longer. Pioneering research was undertaken in the USA, where the California Bearing Ratio (CBR) was developed. The CBR measures the load-carrying ability of the subsoil and is an important criterion for the design of long-lived durable highway pavements, predicting the elastic life and the design of sympathetic tramway tracks (Lesley, 2002).

Conditions for laying new tramway tracks *ab initio* in urban highways are very different from those of the nineteenth century. It is therefore illogical to use nineteenth-century methods when installing track in twenty-first-century roads, resulting in high costs. Criticisms of contemporary tramway track design were recently published (Snowdon, 2004). Various attempts have been made to revise the design of tramway tracks to exploit the strength of highway pavements. One uses shallow precast concrete beams set into the pavement surface. This uses the highway bearing course with a medium CBR as the foundation. This integrates the support of tramcars and heavy road vehicles (Fig. 2.32). A section of LR55 track was installed in the South Yorkshire Supertramway in 1996 (Lesley, 1996) and continues to give maintenance-free service. The LR55 track has been accredited with British Standard BS EN 14811.

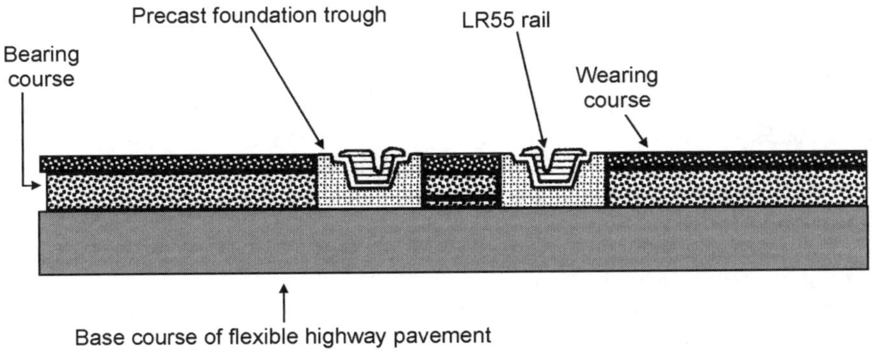

2.32 Integrated tramway and pavement design.

2.11.5 Pavement integrity

The premature failure of highway pavements, where new light rail systems have been installed, is primarily a failure to integrate the loading constraints of tramcars and highway vehicles in the design of the track in the highway. Often the tracks are designed by railway engineers, rather than tramway or highway engineers. More economical and robust tramway installation comes from the integration of design. The interfaces where cyclic loading from road and rail are integral will not cause premature failures. Snowdon (2004) complained that tramway tracks are over-designed and too expensive to build, maintain and modify, but failing to integrate these functional needs.

2.12 Conclusion

This chapter has considered the cyclic excitation mechanisms that can lead to fatigue failure in all the structures that are needed to operate railway vehicles. The different requirements for railways, tramways and light railways were discussed. Some causes of fatigue failures can be removed at the design stage. Many can be reduced. All can be ameliorated or avoided by proper maintenance and planned replacement. By understanding these principles, the cost of building, operating and maintaining railways (and tramways or light rail) can be reduced, while guaranteeing high levels of safety and operational integrity.

2.13 References

Belforte, P., Cervello, S., Collina, A. and Pizzigoni, B. (2006), Numerical and experimental investigation on the effect of resilient wheelset on the rail corrugation phenomenon, *Railway Engineering Conference 2006*, June, London.

Brencich, A. and Colla, C. (2002), The influence of construction technology on the mechanics of masonry railway bridges, *Railway Engineering Conference 2002*, July, London.

Burston, M.C. (2005), A tool to predict rolling contact fatigue, *Railway Engineering Conference 2005*, June, London.

Burston, M., Watson, A. and Beagles, M. (2002), Application of the whole life rail model to control rolling contact fatigue, *Railway Engineering Conference 2002*, July, London.

Cannon, W. and Henderson, E. (2002), Retrofit earthworks to unstable railway embankments, *Railway Engineering Conference 2002*, July, London.

Dembosky, M.D. and Timmis, K. (2005), Rolling contact fatigue – what we have learnt, *Railway Engineering Conference 2005*, June, London.

Dover, A.T. (1965), *Electric Traction*, Pitman, London.

Eisenmann, J. (1970), Stress distribution in the permanent way due to heavy axle loads and high speeds, *Area*, **71**: 24–59.

Fowler, G. (1976), Fatigue crack initiation and propagation in pearlitic rail steels, PhD thesis, University of California.

Harvey, W.T. (2002), Railway arch assessment, *Railway Engineering Conference 2002*, July, London.

Hecht, E. (1994), *Physics*, Brooks/Cole, Pacific Grove, CA.

Howard, M. (2004), *The Rail Engineer*, November, p. 7.

Hughes, M.P. (2002), Yarm Viaduct – foundation stabilisation, *Railway Engineering Conference 2002*, July, London.

Kapoor, A. (2002), Wear/fatigue interaction maintenance strategies, *Railway Engineering Conference 2002*, July, London.

Koc, W. and Palikowska, K. (2002), Railway track design using evolution programming, *Railway Engineering Conference 2002*, July, London.

Lesley, L. (1996), Modelling the behaviour of a new track system when subjected to vibration, *ASME European Structural Dynamics Conference*, July, University of Montpellier, France.

Lesley, L. (2000), Estimating the impacts of rail wagons on new track forms, *ESDA Summer Conference*, July, University of Geneva.

Lesley, L. (2002), Testing the next generation of tramway tracks, *Railway Engineering Conference 2002*, July, London.

Lesley, L. and Al Nageim, H. (1996), Transmission of vibration and conditions of resonance, *ASME Symposium on Structural Reliability*, 31 January, Houston, TX.

Matsuura, A. (1992), Dynamic interaction of vehicle and track, *Quarterly Report, RTRI*, Vol. 33, No. 1, Tokyo.

Morgan, B. (1971), *Railways: Civil Engineering*, Arrow Books, London.

New Civil Engineer (2004), Report, 28 October 2004.

ORE (1966), Stress distribution in the rails, *Bulletin D71*, RP2, Utrecht, The Netherlands.

Orringer, O., Morris, J.M. and Steele, R.K. (1984), Applied research on rail fatigue and fracture in the United States, *Theoretical and Applied Fracture Mechanics*, **1**: 23–49.

Profillidis, V.A. (2000), *Railway Engineering*, Ashgate, Aldershot, UK.

Ringsberg, J.W. and Josefson, B.L. (2002), Modelling of rolling contact fatigue of rails, *Railway Engineering Conference 2002*, July, London.

Robbins, M. (1965), *The Railway Age*, Penguin Books, Harmondsworth, UK.

Savage, R. and Amans, F. (1969), Railway track stability in relation to transverse stresses exerted by rolling stock – a theoretical study of track behaviour, *Rail International*, November.

Snowdon, J.R. (2004), *Tramway and Urban Transit*, November, p. 430.

Timoshenko, S. and Langer, B.F. (1932) Stress in railroad track, *ASME Transactions*, **54**: 277–293.

UIC (1979), *Catalogue of Rail Defects*, UIC, Paris.
Upadhyay, R.K. (2005), Emergence of gauge corner fatigue in rails: understanding and evaluation of causes, *Railway Engineering Conference 2005*, June, London.
Yasojima, Y. and Machii, K. (1965), Residual stress in the rail. *Permanent Way 8*, Japan.
Zarembski, A.M. (1979), Effect of rail section and traffic on rail fatigue life, *American Railway Engineering Association Bulletin 673*, Vol. 80.

3
Fatigue in railway bridges

M. GILBERT, University of Sheffield, UK

Abstract: This chapter reviews fatigue in railway bridges. After a historical overview, it discusses key requirements for railway bridges. It then considers fatigue issues in masonry arch bridges, moving from design and materials to particular types of structural fault such as ring separation, longitudinal cracking and distorted profiles. It also reviews methods for analysing the structural health of this type of bridge. The final sections discuss failure in parapets, metal and concrete bridges.

Key words: fatigue in railway bridges, masonry arch bridges, metal bridges, concrete bridges, parapets

3.1 Introduction

While much of the railway infrastructure in use in the UK in 1850 has long since disappeared (e.g. track, locomotives, signalling), many of the bridges, viaducts and tunnels are still called upon to serve ostensibly their original function. The first part of this chapter thus focuses on how the bridge stock we have now come to inherit has evolved. This situation clearly varies considerably from country to country and the UK situation is focused on here.

A real problem with many older bridges is that understanding how they function structurally can be difficult; hence determining both when intervention is required and what sort of intervention is most appropriate can be similarly difficult. These comments apply particularly to masonry arch bridges, the most common single bridge type on the UK rail network. This structural form is therefore considered in most detail, with tests designed to assess the significance of faults such as ring separation and detached spandrels being described in the chapter. Metal bridges have their own problems and issues such as corrosion and fatigue are briefly considered. Issues affecting reinforced concrete bridges are also briefly touched upon.

Many existing bridge parapets were designed merely to protect livestock and horse-drawn traffic from precipitous drops, yet are now expected to contain road vehicles moving at speed. Masonry parapets are particularly common on overline rail bridges and their ability to protect the railway below is therefore scrutinised and recent research findings discussed.

It should perhaps be stressed that because fatigue failure in metallic structures is already a well-documented scientific phenomenon, it is not considered here in particular detail. Conversely fatigue failure in masonry structures is currently poorly understood and hence in this chapter the primary focus will be on the appraisal of masonry structures which have suffered damage likely to have been caused or at least exacerbated by fatigue, rather than on a detailed analysis of the mechanics of fatigue failure in masonry.

3.2 Historical context

Starting with the Stockton and Darlington Railway in 1825 for goods traffic and the Liverpool to Manchester Railway in 1830 for passengers too, the UK was at the leading edge of the railway revolution. This means that many structures are much older than comparable ones in other countries, and there is consequentially a perennial fear that the UK will suffer first should large numbers of a certain type of structure abruptly reach the end of their working lives. An additional problem is that the UK has a particularly varied stock of bridges. This is because typically each new rail route was instigated by a separate company which often had their own favoured design and construction methods (the UK had over 200 railway companies, compared, for example, with only three main companies in Spain). Additionally, different companies and the engineers they employed adopted different approaches to rail routing, with a consequential major influence on the number and type of structures required. For example, Joseph Locke favoured circuitous routes in hilly terrain (e.g. his West Coast Main Line passes over the Westmorland hills at Shap) whereas others such as Brunel and Stephenson preferred more direct routes, inevitably involving extensive tunnelling and bridge work.

Furthermore, since the UK was at the forefront of developments in the early nineteenth century, many railway structures are now considered to be important parts of the UK's heritage and are now listed or scheduled. Though most would agree that this is a good thing, it does impose limits on the type of engineering interventions permitted in many cases. There have also, for example, been plans for important sites along Brunel's London Paddington to Bristol Temple Meads line to become part of a dispersed UNESCO World Heritage Site which includes the majestic twin-span brickwork arch bridge at Maidenhead and other structures.

In terms of bridge materials, early bridges were constructed using masonry (brick or stone), timber or cast or wrought iron. Timber bridges were particularly favoured for highly speculative routes as they could be constructed relatively quickly and cheaply. However, low durability means that almost all of these have since been replaced with metal and, sometimes, masonry structures.

Although railways allowed the raw materials for new bridges to be transported reasonably long distances, potentially resulting in the use of the same materials for a number of structures on a particular stretch of canal or railway, there is also evidence that large structures often made use of very local materials (e.g. brickworks were installed adjacent to the imposing Ribblehead viaduct on the Settle–Carlisle line).

Cast iron found favour in the late eighteenth century in the construction of mills and factories, and consequently was also often used in early rail bridges. However, design rules for cast iron bridges were tightened following the collapse of Robert Stephenson's Dee Bridge, and subsequently, following the collapse of Inverythan Bridge in 1882, its use for new bridges was prohibited.

Thus wrought iron, which is more ductile and has a much higher tensile strength than cast iron, was widely used instead. Though large pieces of wrought iron could not readily be produced, thin wrought iron plates could be riveted together to form large structural members (e.g. for Robert Stephenson's tubular bridge across the Menai Straits to Anglesey).

In the mid-nineteenth century improved production methods led to steel becoming available as a viable construction material (it had previously been possible to manufacture it only in small quantities, for tools and cutlery, etc.). Though initial uptake was slow, steel was famously used for the Forth Rail Bridge in 1890 and over the following few decades steel replaced wrought iron as the material of choice for metallic rail structures.

Reinforced and subsequently prestressed concrete found favour in the twentieth century, and concrete bridges are used extensively on the rail networks of many countries. However, since in the UK the most intensive period of development of the rail system was in the nineteenth century, comparatively few concrete bridges are in service. Of those that are, many were constructed to replace existing overline masonry arch bridges during the electrification of inter-city lines in the 1960s. Figure 3.1 shows the breakdown of the UK bridge stock

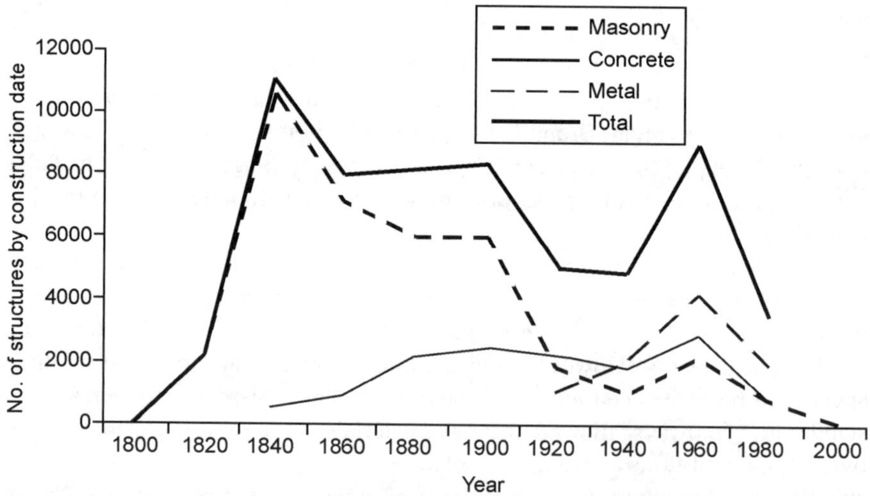

3.1 Breakdown of UK railway bridge types by age and material.[1]

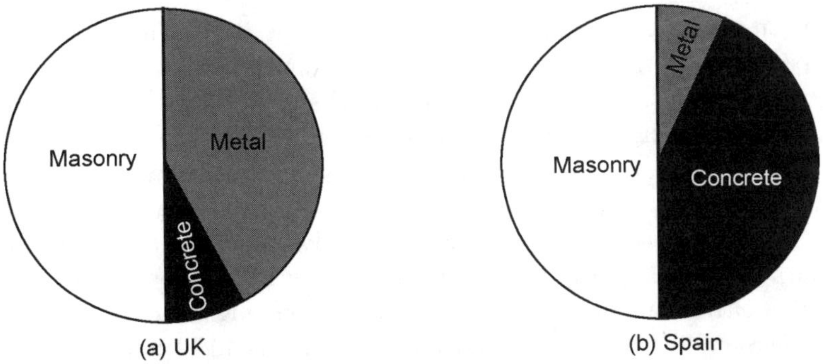

3.2 Comparison of the composition of railway bridge stock in two EU countries.

by age and material.[1] Figure 3.2 shows the difference in the composition of the UK and Spanish railway bridge stocks, showing, for example, the far higher proportion of concrete bridges on Spanish railways.

Figures 3.1 and 3.2 also show the very large numbers of masonry bridges in the UK, and also in Spain (this situation is similar in many other European countries). While masonry arch bridges are undoubtedly potentially long-lived structures, only quite recently are scientific methods of assessing the load-carrying capacities of these bridges becoming available. It is also fair to say that there is considerable scope for improvement of these methods. This issue is particularly relevant for railways, since the main arteries of the highway network (i.e. trunk roads in the UK) do not rely heavily on masonry structures, most of which are now in excess of 100 years old. Because of this, and because the general level of understanding among engineers of the mode of response of masonry arch bridges is quite low, this chapter will focus particularly on common deterioration mechanisms in masonry bridges.

3.3 Railway bridge requirements

3.3.1 Live loading and other requirements

Underline bridges must be capable of carrying a given live train load at a given train speed. Train speed is important because certain dynamic effects are speed dependent (e.g. due to the effects of track irregularities, excitation of the bridge by the rail vehicle, etc.).

UK rail underline bridges are currently not routinely assessed to carry specific trains. Instead the so-called 'Route Availability' system is used, with the standard loading pattern being derived from the weight distribution of an early twentieth century (LNER) steam locomotive. Using the system, 11 units of the standard loading pattern is given the RA number RA1; 12 units is denoted RA2 and so forth up to RA15 (25 units). Because of the importance of train speed, a given bridge could therefore be designated as being suitable for up to RA6 loading at 100 mph and also up to RA10 loading at 60 mph. Since particular train types are allocated an RA category, the system can be used to determine what trains can safely pass down a particular line, and at what speed. In the case of beam bridges the rather complex standard loading pattern can fortunately be simplified to an equivalent uniformly distributed loading.

Loading requirements for overline railway bridges are essentially the same as for any other highway bridge (though the 1968 Transport Act imposes certain additional requirements). Currently maximum highway axle loads are 11.5 tonnes and the maximum gross vehicle weight is 40/44 tonnes.

Additionally, bridges crossing rivers should have substructures which are not prone to scour. In a 1976 study of 143 bridge failures, it was found that 66 of these were due to scour.[2] More scour-induced failures have occurred since; for example, a multispan brick arch railway bridge at Inverness failed in 1989. Thus scour is clearly a real problem and currently the efficacy of various automatic early warning systems is being investigated by bridge owners.

3.3.2 Design life

Currently the notional design life for a new bridge in the UK is 120 years. Network Rail has recently reported that the average age of its bridges is continuing to increase, and that if present low levels of bridge renewal are maintained in the long term, it would be necessary to keep most bridges fit for purpose for about four or five times longer than this nominal 120-year design life.[1]

Of course, different types of bridges are likely to have different natural lifespans, and it is thus inevitable that the average age of masonry arch bridges, for example, will continue to rise (as virtually none are built at present). This is not necessarily a particular cause for concern. However, it may be so in the case of metal bridges, since these are generally less long lived. Additionally, the effectiveness of current (and past) basic repair and maintenance procedures is important when considering the expected life of bridges.

3.4 Masonry arch bridges

3.4.1 Background

The particular importance of masonry bridges on the rail network (e.g. see Fig. 3.2) means that at the time of writing developing a better understanding of masonry bridges is currently a high R&D priority for Network Rail. Particularly important is the development of improved analysis and assessment methods. Once these emerge, they can also inform the appraisal of alternative rehabilitation and repair strategies.

Masonry arch bridges are in many ways very different from the modern steel

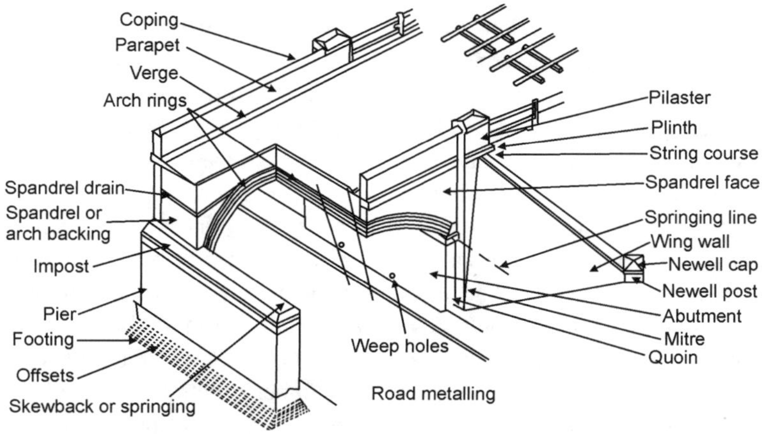

3.3 Masonry arch bridge nomenclature.

and concrete bridges which would typically be constructed instead today. Hence this section first introduces the reader to the design and construction philosophy used in their design, then considers the materials used and their response under load, before moving on to consider some of the structural faults commonly encountered by assessment engineers today, together with potential remedies.

As the terminology used to describe different parts of masonry arch bridges can appear obscure to the non-specialist, common terms are given in Fig. 3.3.

Design and construction philosophy

The vast majority of masonry arch bridges on the rail network appear to have been geometrically sized using empirically rather than scientifically derived formulae. For example, Molesworth's formulae[3] for 'small railway arch bridges' prescribes rise = span/5, ring thickness = span/18 and pier thickness = span/6 to span/7.

Furthermore, it seems that many railway companies had only several standard designs for their bridges, a whole stretch of line being built from these designs.[4] However, conversely for longer spans, recourse did have to be made to scientific methods (e.g. in the case of Brunel's Maidenhead Bridge[5]).

The design of multi-span bridges was revolutionised by Perronet in the mid-eighteenth century, and railway engineers took full advantage of this. Perronet had realised that the thrusts from adjacent spans would counterbalance each other, the consequence being that relatively slender piers could successfully be used, giving rise to more economical and elegant structures. However, the use of slender piers also had its drawbacks, most notably that several spans of a given bridge normally had to be constructed together. Additionally, after construction, removal of one span would cause the entire structure to collapse. This effectively occurred with tragic consequences in 1992, when the centre span of the bridge at St Johns station, London, was being demolished.[6] Numerous other accounts of similar incidents in which adjacent spans collapsed following the removal of a single span can be obtained from around the UK. Thus bridges containing many spans often also included integral 'kingpiers' (very substantial piers) at intervals both to make construction easier and also to prevent the entire viaduct from collapsing in the event of the accidental removal of a single span.

In order to reduce the time and cost of building an arch, many nineteenth century arch bridges were built of brickwork, rather than stone masonry (better mortars and more consistent bricks were being developed, and in the heyday of railway building there would have been a shortage of skilled stonemasons).

Unlike traditional stone voussoir arches (Fig. 3.4a), the barrels of brickwork arch bridges were often built up using a number of 'stretcher' bonded arch rings (Fig. 3.4b). Alternatively, some arch barrels were built using a bond including 'headers' interconnecting adjacent sections of brickwork (Fig. 3.4c). Other barrels used a combination of these patterns, being built up from a number of thick rings of header bonded brickwork, but with no (brick) bonding between these thick rings (Fig. 3.4d).

However, because the use of header bonded brickwork in arches generally requires that specially manufactured tapered bricks are used, the use of stretcher bond was generally more popular, the latter bond having no headers connecting adjacent sections of brickwork. Unfortunately it may be suggested that all stretcher bonded brickwork arches are defective, whether or not adjacent rings appear to have become separated. This is because the inter-ring mortar joints form potential weak surfaces.

Old records provide clues as to the thinking on this issue of the engineers at the time. For example, it is on record that Robert Stephenson and Isambard Kingdom Brunel were in dispute over how best to treat stretcher bonded brickwork arches.[7] Whereas Brunel optimistically suggested that a multi-ring brickwork arch barrel should be treated 'as a homogeneous, and, he might almost say, an elastic mass', Stephenson, perhaps characteristically, was more cautious, suggesting that 'The arch, *per se*, should always be considered as composed of separate masses, not set in a [cement/mortar] matrix, but combined in a certain form, the only adhesion being the friction of the surfaces'. This was

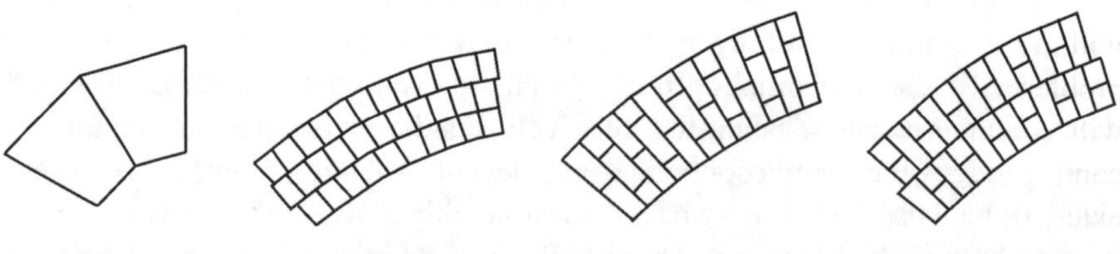

(a) Stone voussoir (b) Brick: stretcher (c) Brick: header (d) Brick: header/stretcher

3.4 Typical masonry arch bonding patterns.

probably also the opinion of G.P. Bidder, later to follow his eminent colleagues as president of the Institution of Civil Engineers. Bidder suggested that 'the line [of thrust in brick arches] would in almost every instance, travel out of the ring in which it commenced, and in case of fracture, the rings would fail consecutively', unless the arch was 'well bonded together throughout its entire depth'. It appears therefore that engineers were aware of the issue, though this did not stop the practice of constructing stretcher bonded arch barrels, perhaps because stronger mortars were being gradually introduced. This issue will be discussed in more detail in Section 3.4.3.

In the majority of relatively short, single-span bridges, spandrel walls were built at the edges of the barrel, and the resulting spandrel void was backfilled to provide a level surface for the road or railway. There is a good deal of evidence to suggest that the restraint afforded by the backfill material at either end of the span will significantly increase the load-carrying capacity of a given bridge. There is also evidence to suggest that the spandrel walls will often be able to provide additional restraint to the arch barrel under loading. These important points will be raised again later.

In order to reduce the dead weight of many long-span or multi-span bridges, the spandrel voids were often not backfilled. Instead internal spandrel walls (sometimes called longitudinal or 'sleeper' walls) were constructed to transfer applied loads onto the arch barrel.[8] In the case of multi-span bridges these walls are likely to have the very important additional effect of propping apart the barrels of adjacent spans.

Despite the emergence of other materials, numerous bridges constructed using masonry continued to be built in the late nineteenth century and many in the early twentieth century. For example, the Ribblehead Viaduct, one of the most well-known masonry arch bridges in Britain and now a scheduled ancient monument, was constructed in 1875, while the Nairn Viaduct over Culloden Moor, the longest masonry arch bridge in Scotland and built of red sandstone, was not completed until 1898.

Resistance to applied loads

It was pointed out by Hooke in the seventeenth century that 'as hangs a flexible line, so, but inverted, stands the masonry arch'. Thus a simple hanging cable will, once frozen, carry its own weight if inverted (this ignores stability problems). If the inverted cable is symmetrically increased in thickness then it will continue to carry its own self-weight without inducing any bending (the compressive force, or 'thrust' simply increasing). Clearly the application of an external load will induce bending which can be idealised internally as an eccentric thrust. Provided this thrust line everywhere lies within a voussoir arch, then the arch will remain stable. It can also be shown that there is no unique thrust line associated with a stable arch – there are many possibilities.

So how does a voussoir arch fail when overloaded? From what has already been said, it follows that if the load is increased such that no possible line of thrust can be found to lie everywhere within the masonry, then the arch will become unstable and fail, often in a hinged failure mechanism. Short- and medium-span arch bridges are most prone to such failures, whereas in the case of long-span bridges foreseeable live loads will generally be relatively small in comparison to dead load effects (and then, from an assessment engineer's perspective, weathering and long-term creep effects become governing factors affecting bridge integrity).

3.4.2 Materials used in masonry arch bridges

Masonry is a composite material comprising masonry units bonded together using mortar. For the units, as well as traditional natural stone and clay bricks, precast concrete blocks were also sometimes used in the UK, most notably on the Glenfinnan viaduct in Scotland.

The range of mortars available to engineers increased significantly in the nineteenth century. In addition to traditional hydraulic lime mortars, in the early part of the century so-called 'Roman' cement was quite widely used (e.g. in Brunel's Maidenhead Bridge, 1839). One of its main characteristics was that it was slow setting. Modern Portland cement was invented by Joseph Aspdin in 1824. This was so named because it was supposed to resemble Portland stone when it set. Portland cement production techniques were then refined in the 1850s. The ratio of the amount of cement to sand used governs the strength of the mortar. In masonry it is usual, and normally desirable, that the mortar used has a very much lower compressive strength than that of the masonry units. This allows movement to be accommodated by comparatively flexible mortar joints.

It is often assumed that stresses in masonry gravity structures such as arch bridges and retaining walls are very low, and hence that material response is of little importance (i.e. because failure will predominantly involve rigid-body deformations). In fact the extent to which this is true depends on several things, not least the span being considered and the form of construction of the arch (e.g. whether or not this comprises multiple stretcher bonded arch rings; see Fig. 3.4).

To illustrate the influence of span it is perhaps instructive to consider the gravity-induced stresses in two geometrically similar masonry arch bridges, say of span L and xL. The scaling law for stresses means that the stresses in the latter bridge will be x times those in the former.

Mechanical performance of masonry

Fundamentally both masonry units (whether brick, concrete or rock) and mortar are quasi-brittle materials whose mechanical performance will deteriorate (soften) under monotonic or cyclic loading. Under modest applied stresses, any micro-cracks in the masonry can be assumed to be stable. However, above a certain stress level, micro-cracks will become unstable. This means, for example, that severe deterioration can eventually occur when stresses of moderate intensity are applied cyclically ('fatigue' failure) or applied statically over a very long period of time ('creep' failure). Creep failure is normally supposed to be the root cause of the sudden failures of various ancient masonry Italian churches (e.g. St Marks, Venice in 1902; Pavia in 1989).

Typical stress–displacement graphs for quasi-brittle masonry materials are shown in Fig. 3.5. Note that usually the compressive strength is at least one or two orders of magnitude greater than the tensile strength, which is small or may even be non-existent. In compression, masonry normally fails due to the formation of tensile cracks parallel to the direction of the applied load.

Additionally, masonry materials can of course degrade due to environmental factors (e.g. chemical and frost damage, etc.) rather than simply due to stress-level related effects.

Backfill material

To date, studies of the performance of masonry arch bridges have tended to focus predominantly on the masonry parts rather than on the backfill material present in the spandrel void zone (assuming the latter does contain backfill, rather than longitudinal walls). However, as noted previously, the ability of backfill material to both disperse the applied loading and offer horizontal restraint to sway of an arch can dramatically enhance bridge strength.

A recent study of highway bridges by the author has revealed wide variations in the backfill used in apparently similar bridges. This is also likely to be true on the rail network. Additionally, in the study much of the backfill was found to be cohesive (i.e. clayey) rather than purely frictional (i.e. granular). Thus in order to obtain a realistic estimate of bridge strength it may in future prove to be desirable to include both cohesive and frictional strengths in the analysis.

3.4.3 Structural fault: ring separation in multi-ring brickwork arch bridges

Though brickwork was used quite extensively in other countries (e.g. 27% of the masonry rail bridges in Spain are brickwork), it seems that the use of the stretcher bonding pattern in the arch barrel is very much more common in the UK than in other parts of Europe (e.g. virtually no rail bridges in Spain use the stretcher bonding pattern). Stretcher bonded arch barrels are problematic because the mortar joints between adjacent stretcher bonded rings form potential surfaces of weakness, with the micro-cracks inevitably present initially in the joints between rings likely to grow under the action of cyclic, fatigue-type loads.

The traditional means of identifying whether or not adjacent stretcher bonded

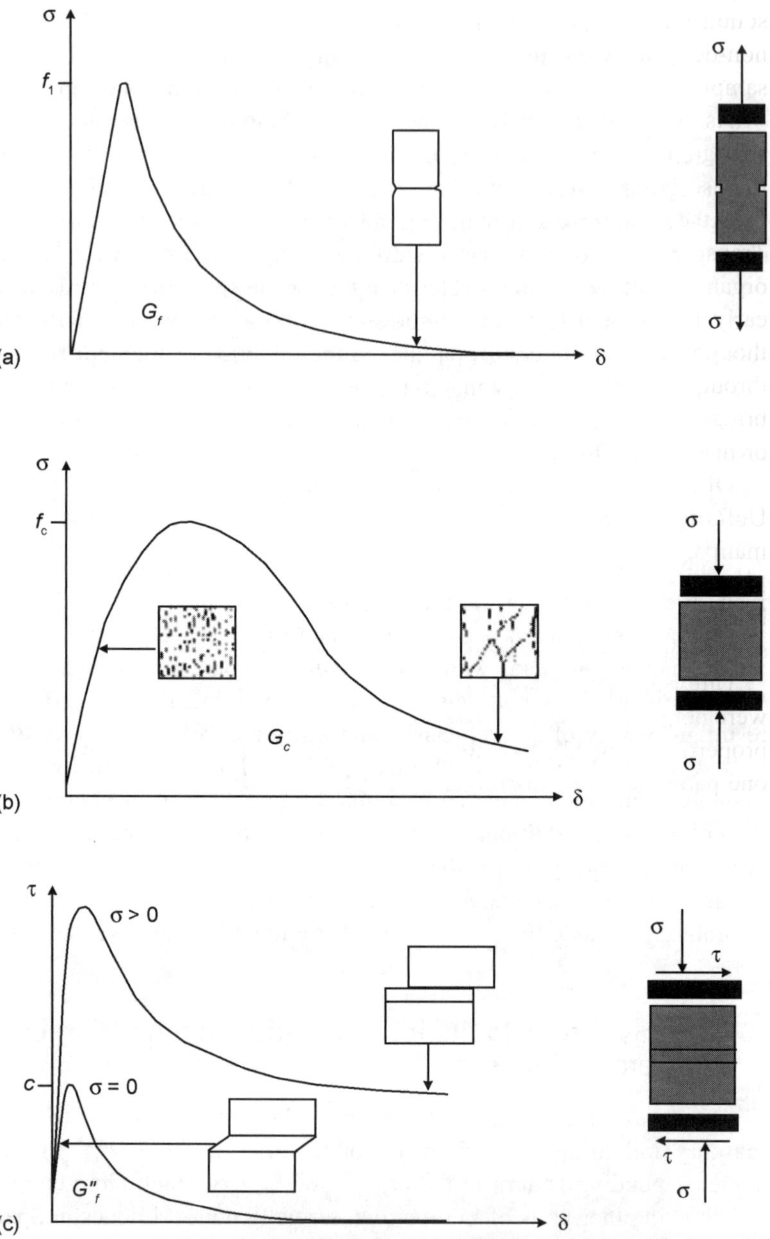

3.5 Typical behaviour of quasi-brittle materials under (a) uniaxial tensile, (b) compressive and (c) shear loading (after Lourenco, P.B., *Computational strategies for masonry structures*, Delft University Press, Delft, 1996).

rings have become debonded is to tap the bottom ring with a hammer; a hollow sound indicates physical separation of the bottom ring. Alternatively, modern non-destructive testing (NDT) techniques can be used or traditional cored samples taken and inspected for evidence of ring separation.

As far as engineers are concerned, an important question is how bridge strength is affected by the presence of ring separation, or indeed more generally the use of brickwork rather than stonework to form the arch barrel. To investigate the comparative strengths of both brickwork and stonework arches, the Transport and Road Research Laboratory (TRRL, now TRL) in the UK organised a programme of tests on redundant arch bridges in the late 1980s and early 1990s.[9] Most bridges tested were found to fail in four hinge mechanisms, though some of the bridges were reported as failing by 'three hinge snap through' or in 'compression' (material failure). It was likely that many of the bridges tested were restrained considerably by their attached spandrel walls and/or masonry backing.

Of the bridges tested, Torksey Bridge contained a multi-ring brickwork arch. Unfortunately, the spandrel walls of this bridge had become detached from the main arch barrel and these obscured the view of the arch rings prior to collapse. However, in the case of Rotherham Road bridge,[10] a three-span overline railway bridge which comprised a mixed header–stretcher bonded arch barrel, ring separation was clearly evident (Fig. 3.6).

Unfortunately, detailed pre- and post-test investigations of the field bridges were not performed (i.e. to determine internal constructional details and material properties), which means that it is quite difficult to isolate the influence of any one parameter, such as ring separation. In the case of Rotherham Road Bridge,

3.6 Overline rail bridge being load tested to collapse: ring separation highlighted.[10]

strong backing material in the spandrel void area between spans was clearly visible following the test, the latter masking any reduction in carrying capacity caused by the onset of ring separation.

Therefore, to investigate the particular fault of ring separation further, in the early 1990s British Rail Research funded a series of laboratory tests on multi-ring brickwork arches,[11] with the main aim being to determine the reduction in carrying capacity which would result from the presence of ring separation, or 'delamination'. The tests provided very significant findings, and hence further details of the tests are included here. Both arch ribs and arch bridges were tested. The barrels of the arch bridges to be described here were not extended underneath the spandrel walls, with the consequence that the spandrel walls were effectively detached (see Section 3.4.4), leaving the arch barrel free to slide past the spandrel walls. This simplifies the structural system. Also, for consistency, all arch ribs and arch bridges had segmental profiles with span to rise ratios of 4:1. Details of three arch ribs and five arch bridges tested are given in Table 3.1 and Fig. 3.7.

Table 3.1 Laboratory bridge test details

(a) arch ribs and bridges tested

Ref.	Span (m)	Comments
arch rib 1	3	–
arch rib 2	3	Inbuilt ring separation*
arch rib 3	3	Inbuilt ring separation* + 5 inter-ring 'header' bricks
bridge 3-1	3	–
bridge 3-2	3	Inbuilt ring separation*
bridge 5-1	5	–
bridge 5-2	5	Inbuilt ring separation*
bridge 5-3	5	Spandrel tie bars used

*Damp sand used in place of standard mortar (in tangential joint between arch rings)

(b) construction materials

Material	Comments
Mortar	1 : 2 : 9 cement : lime : sand (except 1 : 1 : 6 for tangential faces of arch rib 1)
Bricks	Full-size solid class 'A' Engineering (radial faces coated with release agent for arch ribs)
Brickwork	Stretcher bond pattern; density 2310 kg/m^3; brick–mortar joint friction coefficient 0.64 (0.53 in case of damp sand–unit joint friction)
Backfill	50 mm downgraded crushed limestone, compacted in 150 mm layers to a density of 2260 kg/m^3 (arch bridges only)

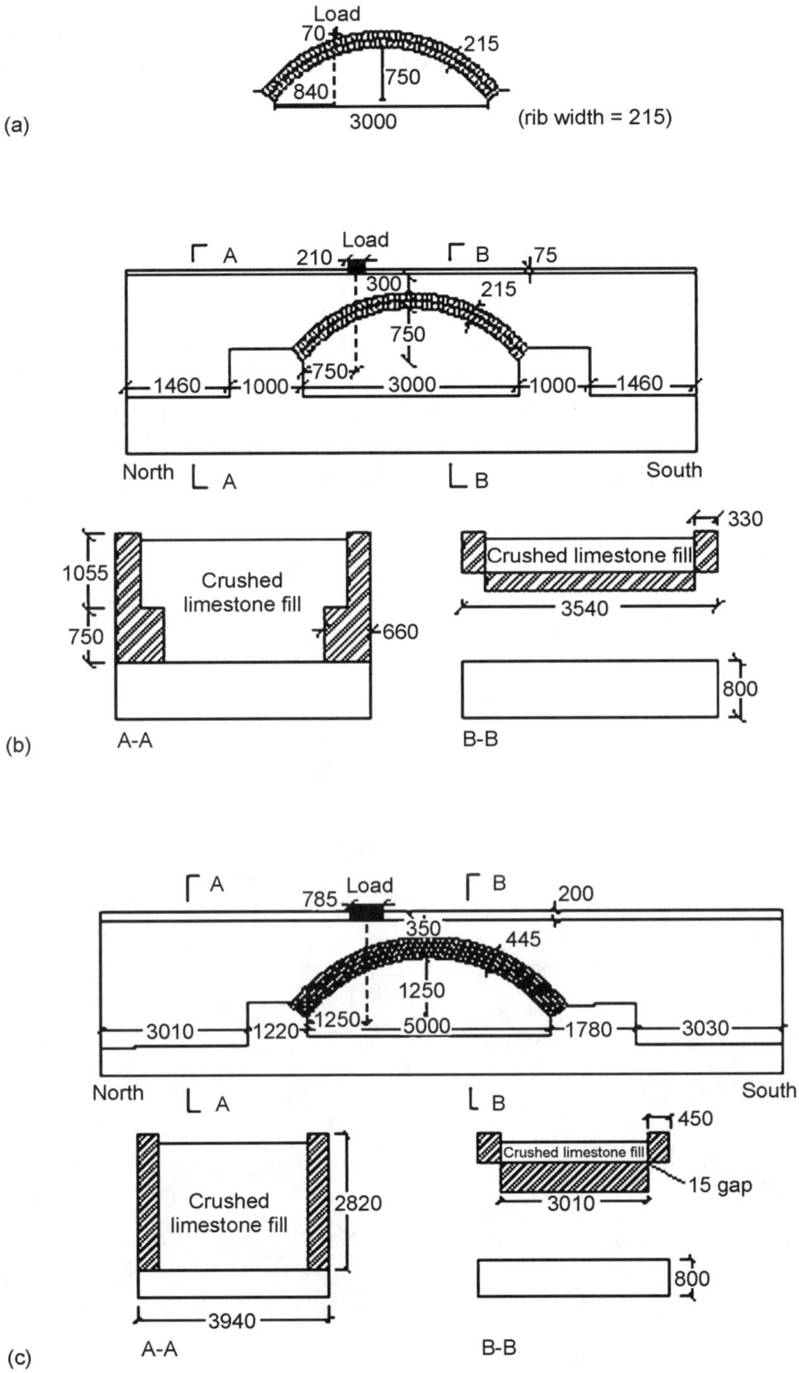

3.7 Ring separation study: laboratory arch ribs and arch bridges (dimensions in mm): 3 m arch ribs; (b) 3 m arch bridges; (c) 5 m arch bridges.

Hydraulic loading systems were used to load test all the arch ribs and arch bridges to ultimate collapse. The resulting failure mechanisms and load deflection plots are given in Figs 3.8 and 3.9 respectively. It is evident that only arch rib 1 and bridge 3-1 failed in a conventional four-hinged mechanism.

The main findings may be summarised as follows:

- Full ring separation leads to very significantly reduced bridge strength. The largest reduction in strength, approximately 70%, was found in the case of the 5 m span bridges comprising four-ring arch barrels.
- Arches with intermittent inter-ring connections (e.g. due to the presence of headers) are likely to be stronger than those with full ring separation but weaker than arches with full inter-ring connectivity.
- Some of the bridges constructed with the mortar-bonded arch barrels failed unpredictably and abruptly, developing partial ring separation and consequentially suffering sudden reductions in carrying capacity.

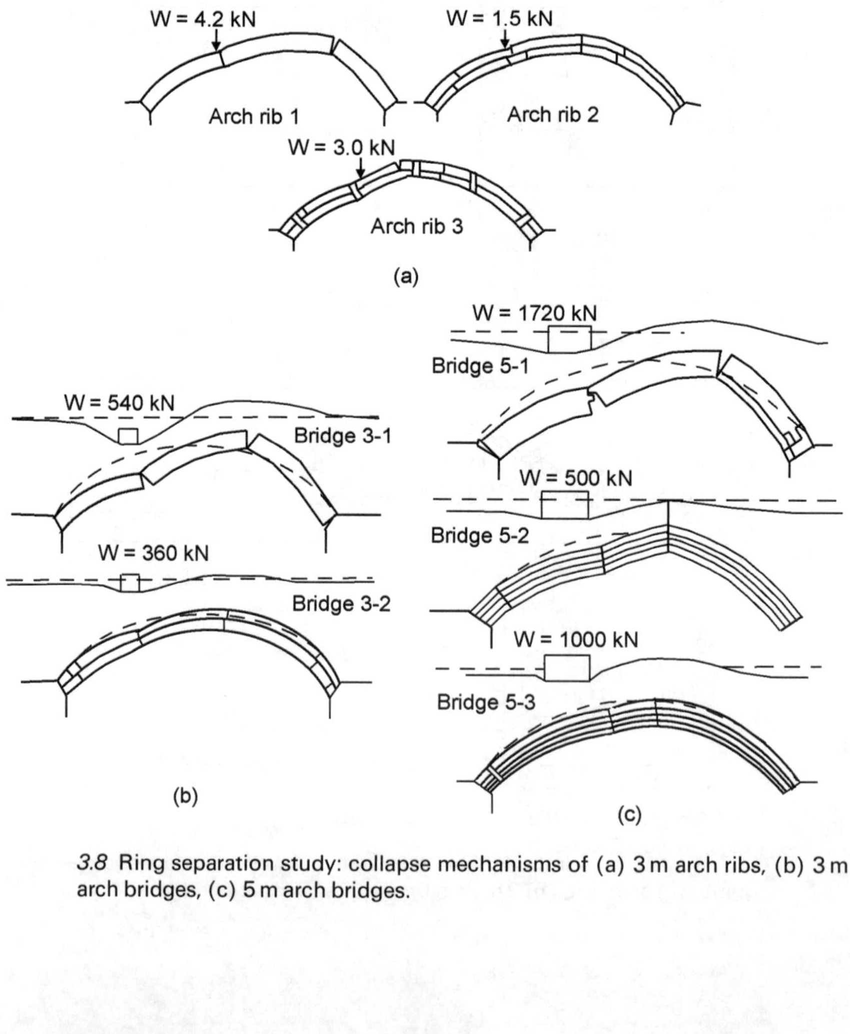

3.8 Ring separation study: collapse mechanisms of (a) 3 m arch ribs, (b) 3 m arch bridges, (c) 5 m arch bridges.

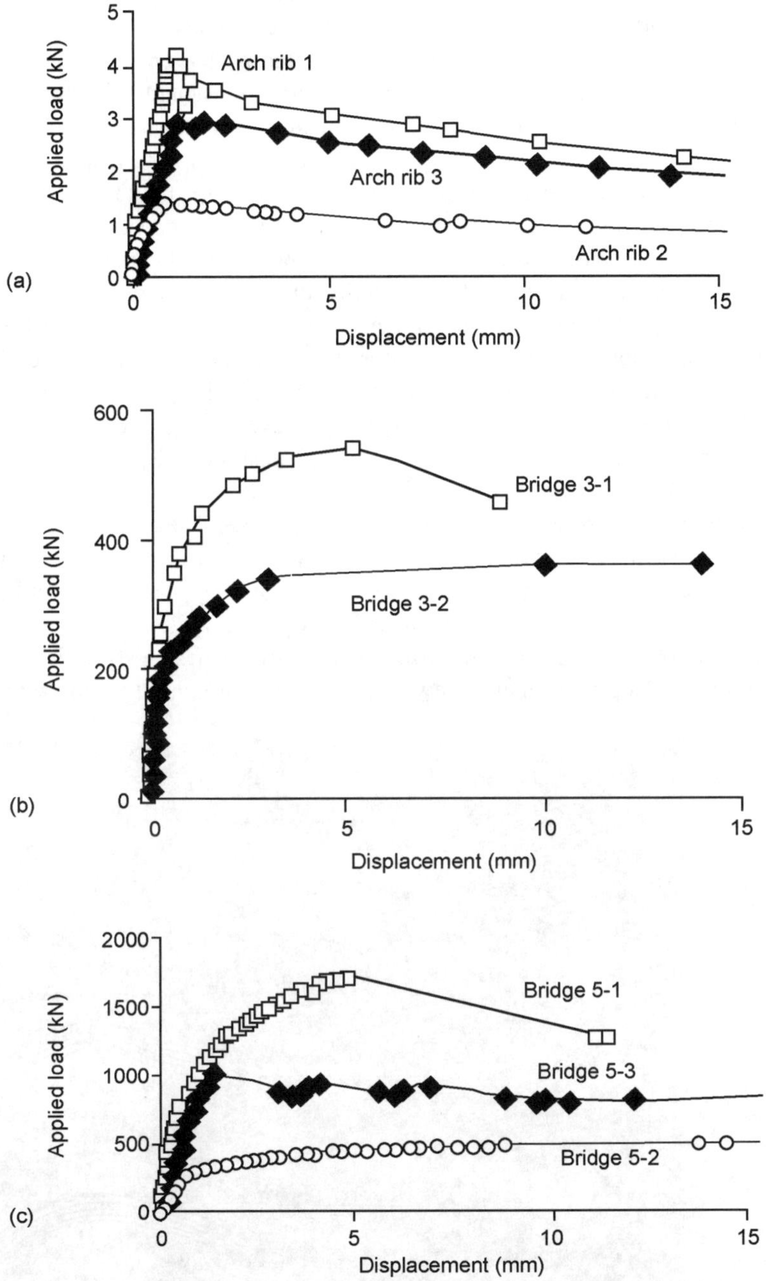

3.9 Ring separation study: load vs. displacement responses of (a) 3 m arch ribs, (b) 3 m arch bridges, (c) 5 m arch bridges.

On the last point, it should be emphasised that two bridges (5-1 and 5-3) were nominally identical yet failed at very different loads: 1720 kN and 1000 kN respectively. In both cases failure was initiated by the onset of ring separation. In the case of bridge 5-3 the post-failure 'saw-tooth' load–displacement relationship is due to the incremental spread of ring separation.

While the use of stronger mortar would undoubtedly have helped to reduce the possibility of ring separation in the laboratory tests, it should be noted that with increasing spans the scaling law for stresses in gravity structures means that stresses will quickly exceed those which can be resisted by any traditional mortar (i.e. ring separation is inevitable). Additionally, in the case of bridges built before the development of Portland cement, the strength of the 1 : 2 : 9 (cement : lime : sand) mortar used in the laboratory tests may, if anything, represent an upper bound on the strength of that which was used originally (ignoring any weathering and fatigue-related deterioration which may have taken place since).

In the case of the arches with inbuilt ring separation, designed to replicate the condition of real structures which have 'shaken down' to a stable state, the load–deflection response exhibited a long ductile plateau. The response of these latter bridges was quite different from that of voussoir arch bridges in that multiple small hinges actually formed (Fig. 3.10; only the major hinges are shown in Fig. 3.8).

Additionally, a number of skew arch bridges of comparable geometry to the 3 m span bridges described have now been tested. These tests indicated that the 3D load paths present make skew bridges even more prone to ring separation.[12]

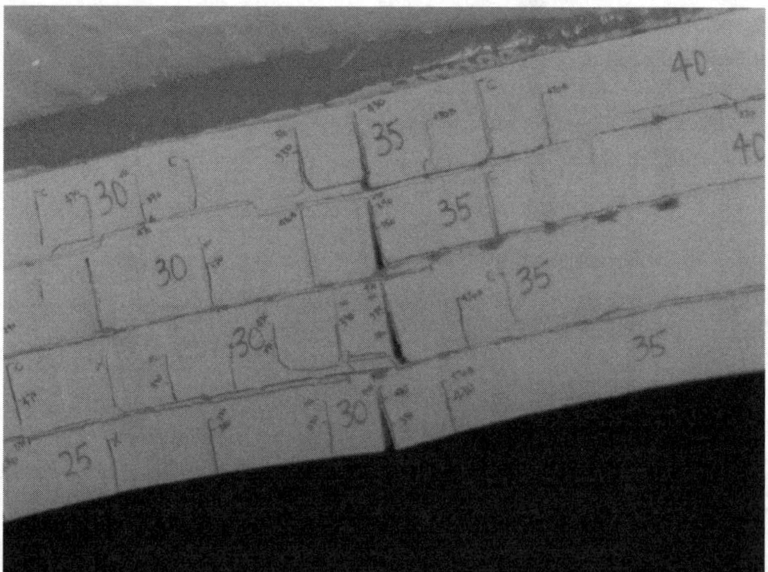

3.10 Diffused radial cracks in a laboratory multi-ring brickwork arch bridge (bridge 5-2).[11]

Furthermore, others have recently undertaken centrifuge testing of very small-scale model multi-ring brickwork arch bridges, and have frequently identified ring separation.[13] However, it is not yet clear that centrifuge tests are a suitable means of replicating the brittle fracture failure modes encountered in large-scale structures, so it is difficult to draw firm conclusions.

Laboratory load tests (cyclic)

Clearly, real bridges in the field are not loaded monotonically to collapse. Significantly, there is a possibility that under cyclic loading damage may occur at a much lower load than in a monotonic load test. Conclusions from recent compression tests on brick masonry specimens were that the fatigue strength of dry brick masonry at 10^8 cycles was approximately 50% of its static strength under comparable loading conditions, and that saturated masonry was found to have considerably reduced fatigue strength.[14] However, it was also found that the compressive strength of masonry specimens subjected to the sort of non-uniform (i.e. eccentric) loading found in masonry arches was considerably greater than its corresponding strength under uniform loading. Furthermore, it must be borne in mind that in the case of many masonry arch bridges the compressive strength of the masonry is of secondary importance in determining ultimate bridge strength.

Additionally, cyclic fatigue load tests on multi-ring arches have been conducted at Nottingham University and more recently at Salford University.[15] As might be expected, it has been found that inter-ring cracks will propagate under the action of repeated loads, leading in due course to a bridge with significantly reduced load-carrying capacity. Researchers from Salford University have, for example, proposed that an endurance limit surface for multi-ring brickwork arch bridges may be suitable (Fig. 3.11).

Some research on the fatigue performance of strengthened multi-ring arches has also been performed. An interesting finding was that provision of intrados reinforcement can actually reduce the load-carrying capacity of multi-ring masonry arches, by changing their mode of failure.[16] However, although some basic principles can be transferred from the more widely researched field of concrete fatigue, it is clear that considerable further research is still required in order to increase our general level of understanding of the performance of masonry and masonry structures under cyclic loading.

3.4.4 Structural fault: longitudinal cracking in masonry arch bridges

Longitudinal cracks are commonly found either directly underneath the spandrel walls ('spandrel wall detachment') or near the centreline of an underline bridge when the latter carries twin tracks and trains in both directions. Unlike the fault

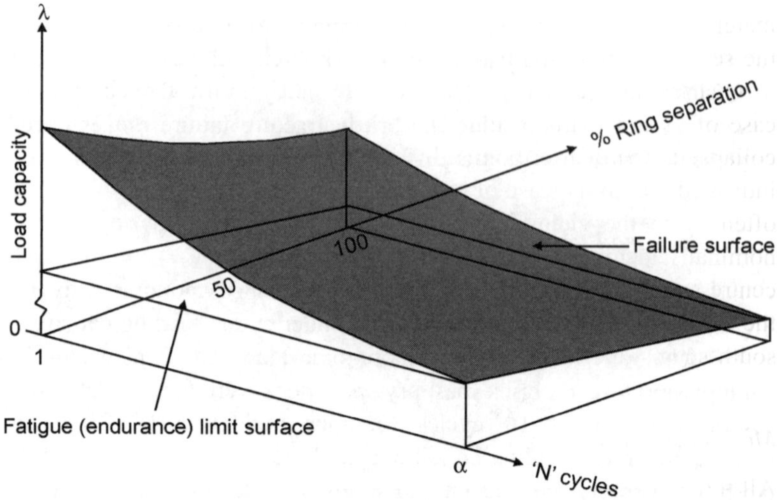

3.11 Proposed endurance limit surface for brickwork arch bridges.

of ring separation, the presence of interlocking masonry units means that longitudinal cracks normally pass both through masonry units and through joints (rather than just through the joints). Like ring separation cracks, these are likely to gradually propagate under the effects of cyclic loading.

Laboratory tests have been conducted to investigate the influence of spandrel wall detachment on the load-carrying capacity of arch bridges.* Tests on single-span bridges[11] indicated that spandrel wall detachment may reduce bridge stiffness, but may have little influence on carrying capacity (however, this latter finding is perhaps unsurprising since in the tests the ends of the spandrel/wing walls were unrestrained). Additionally, tests on multi-span bridges[17] variously containing attached and detached spandrel walls were also conducted and, because of the ubiquity of multi-span bridges on the rail network, further details of the three tests performed will be presented here.

Each test bridge contained three spans, with geometries identical to those of the 3 m span single-span bridges described previously. Multi-span 1 was built with attached spandrel walls, whereas multi-span 2 was built with walls which were detached from the arch barrels of each of the spans, but which were constructed on the same piers and abutments supporting the arch barrels. The intermediate piers of the bridges were designed to have similar ratios of pier height to pier thickness and of arch span to pier thickness to those used in practice during the nineteenth century (pier height 1500 mm, width 440 mm). The wing and retaining walls of the bridges were brick bonded together. The

* It should also be noted that there is a danger of a detached spandrel becoming unstable and simply 'falling off' a bridge.

materials used were from the same sources as those used for the construction of the series of single-span arch bridges described previously.

Quarter span is usually considered to be the critical loading position in the case of masonry arch bridges, and hence multi-spans 1 and 2 were loaded to collapse at the quarter point of the centre span. However, mechanism analyses indicated that in the case of multi-span bridges the critical loading position may often be in the vicinity of the crown. Thus a further bridge (multi-span 3), nominally identical to multi-span 1, was load tested 240 mm to the north of the centre span. Concrete blocks (total weight 48 kN) were placed at the crown of the north span to ensure that this bridge failed in a mechanism involving the south span, which had been more heavily instrumented.

Modes of failure

All three bridges failed in hinged mechanisms. Figure 3.12 shows the modes of failure of the bridges. The load vs. displacement responses for the bridges are given in Fig. 3.13, in which the radial displacement was measured at the quarter span (multi-span 1, 2) or the crown (multi-span 3) of the centre span, on the bridge centreline.

The main findings may be summarised as follows:

- Spandrel walls have the potential to significantly stiffen and strengthen multi-span masonry arch bridges (e.g. compare the responses of multi-spans 1 and 2).

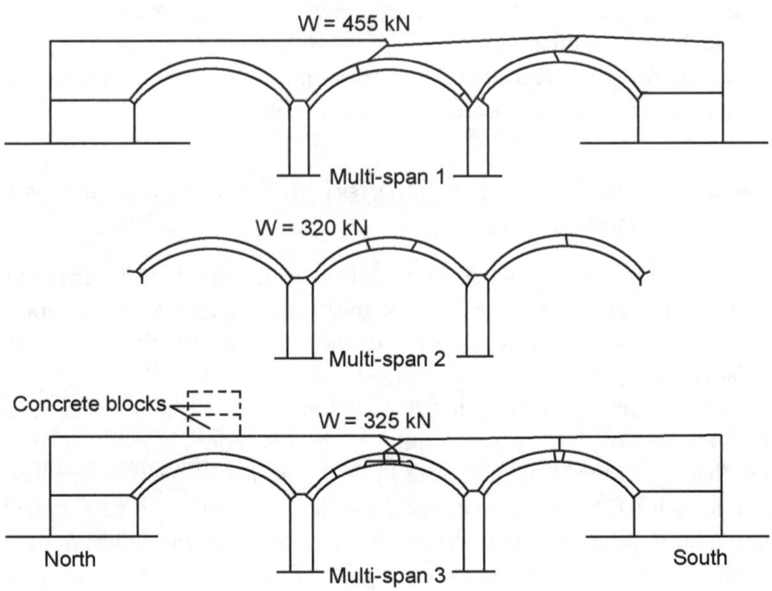

3.12 Spandrel wall detachment study (multi-span bridges): collapse loads and modes.

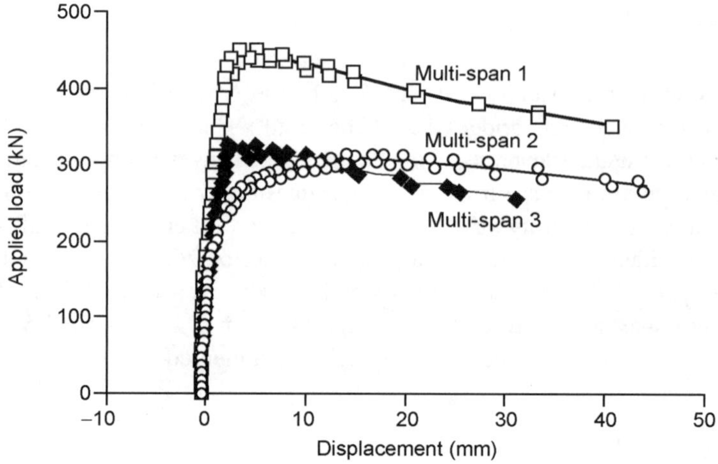

3.13 Load vs. displacement response of multi-span arch bridges.

- The critical loading position for multi-span bridges is usually in the vicinity of the crown.
- Once again some ring separation was observed (under the load in the case of multi-span 3).

The implications of the results for bridges in the field are that bridges with detached spandrel walls will be more flexible and may well also have a lower ultimate strength than their counterparts with attached spandrels.

Finally, it is indicative of the comparative lack of research in the field that these are the first recorded large-scale multi-span arch bridge tests in the literature, yet were conducted only in the 1990s.

3.4.5 Structural fault: distorted profiles in arch bridges and tunnel linings

Another common fault is that of distortion of the arch barrel. Since the shape of an arch in relation to the pattern of the applied load governs overall stability, when the shape becomes distorted the stability of the arch can become endangered.

Consider first distortion that does not arise from spreading or settlement of the supports. In this case the cause of the distortion is likely to be persistent overloading of the structure, perhaps over a period of decades or more. Bridges that have suffered ring separation and/or spandrel wall detachment will be particularly prone to such distortion, and thus fatigue-induced damage may perhaps be regarded as the initial trigger of this fault. To obtain a feel for the influence of the magnitude of the distortion on carrying capacity, it is useful to refer back to Fig. 3.9, which shows the load–deflection responses of the single-span bridges tested to collapse. It is evident from the figure that arches can be quite sensitive to relatively small deformations.

It is also evident from the figure that multi-ring arch bridges containing inbuilt ring separation appear the least sensitive to distortions. However, as a note of caution, because much of the resistance of these structures is due to friction (relative sliding between rings) rather than due to geometrical stability attributable to eccentricity of the hinges, collapse when it does occur is likely to be very sudden.

Alternatively, very often the distortion is caused by settlement of one of the foundations, with the impact on load-carrying capacity being dependent on the type and magnitude of the settlement (i.e. horizontal, vertical and/or rotational).

In the case of brickwork tunnel linings, ground movements, either ongoing or perhaps caused by nearby construction work, may cause the tunnel profile to change. In this case the considerations are (i) whether the stability of the lining has become endangered, and (ii) whether there is still adequate clearance for rail vehicles within the distorted tunnel. Unfortunately, assessing the stability of such tunnels can be very difficult, since not only is the integrity of the bond between brickwork rings not known, the intensity and distribution of the soil pressures are also uncertain (though some indirect evidence can be gleaned from the form of the distorted shape). In practice, one pragmatic solution is to heavily instrument such a tunnel lining with crack sensors, with these triggering inspection or even line closure in extreme cases.

3.4.6 Assessment and analysis techniques for masonry arch bridges

The MEXE (Military Engineering Experimental Establishment) method of assessment is at the time of writing the most commonly used method for both railway and highway bridges in the UK. Developed during the Second World War, the method was initially designed as a quick and easy assessment tool for use by the military, before later being adapted for civilian use. The method is loosely based on a limiting stress analysis of a centrally loaded two-pin elastic parabolic arch. A series of modifying factors have to be applied in order to allow the method to be applied to arch bridges with varying profiles, span to rise ratios, materials, etc. Finally, a global condition factor is applied which allows an assessment engineer to include a subjective assessment of the impact of distortion, cracks, etc., on carrying capacity.

It is commonly stated that the method is conservative, except for long spans. In fact there is increasing evidence that the MEXE method will often be found to give non-conservative assessments for many short-span bridges. The problem is that the structural idealisation, assumed loading position and permissible stress failure criterion used in the formulation of the MEXE method are basically inappropriate (e.g. the critical loading position for a single-span bridge is actually likely to be near the quarter span, rather than at midspan). Perhaps the main reason the MEXE method has continued to be used over such a long period is connected to the fact that masonry arch bridges tend to deteriorate relatively slowly even when repeatedly overloaded. Hence, providing the bridge is regularly inspected and assessed, signs of deterioration will lead to a reduced condition factor and consequentially eventually to a more reasonable assessment of the strength of the structure. However, as a predictive tool (e.g. to ascertain prior to the event whether or not a given bridge is capable of carrying increased vehicle axle weights without being damaged), unfortunately MEXE cannot be expected to help. This means that the method cannot be expected to play a useful part in sophisticated bridge management systems of the future.

Mechanism analysis

In its simplest form an arch bridge may be considered as a two-dimensional assemblage of wedge-shaped voussoir stones spanning between unyielding supports and subjected to vertical loading. Pippard and Baker[18] showed that a common mechanism involved the formation of four 'hinges', and Heyman[19] (and others) then applied plastic theorems to develop the 'mechanism' methods of analysis which now form the basis of several commercial analysis programs.

The basic 'mechanism' method assumptions were that: (i) the masonry in the arch has no tensile strength; (ii) the masonry in the arch is incompressible; and (iii) sliding between the masonry units cannot occur. Despite these simplifications, the 'mechanism' method provides a useful means of rapidly estimating bridge strength prior to carrying out a more detailed and, perhaps arguably, more accurate analysis.

Rigid block mechanism analysis

Because of the no-sliding assumption, the basic mechanism method is clearly not appropriate for analysing multi-ring arch bridges (as inter-ring sliding might occur). Fortunately, Livesley[20] presented a general computer-based 'rigid block' mechanism analysis formulation which allowed the no-sliding and infinite compressive strength assumptions to be discarded. The same basic method was later applied by the present author to multi-ring arches and arch bridges described previously.[11] This allows multi-ring arches to be modelled simply as assemblages of discrete blocks.

For example, consider the analysis of the laboratory model arch ribs and arch bridges described in Sections 3.4.3 and 3.4.4, with all model parameters taken as the measured values given in Table 3.1 and Fig. 3.7. Predicted strengths of arch ribs 1, 2 and 3 were respectively 3.84 kN, 1.42 kN and 3.24 kN, which are reasonably close to the experimental values of 4.2 kN, 1.5 kN and 3.0 kN.

However, the analysis of the model arch bridges is somewhat complicated by the interaction of the arch and backfill material. To appreciate the significant influence of the latter, consider for example bridge 3-1. If the backfill is assumed merely to act as vertical dead weight, then the predicted rigid block analysis failure load for this bridge is 165 kN (just 30% of the test value of 540 kN). However, introducing horizontal backfill pressures as measured using pressure cells positioned in the fill (average pressure of 55 kN/m^2 for the 3 m span bridges) and also accounting for dispersal of the applied load, leads to a dramatic increase in the predicted failure load, to 553 kN.

Using the same backfill modelling approach for ring-separated bridge 5-2 (though with average backfill pressures of 90 kN/m^2 for the 5 m span bridges) produced a predicted carrying capacity of 468 kN. This compares reasonably well with the observed value of 500 kN. The rigid block analysis model can also be used in cases where spandrel walls are attached; e.g. consider multi-span bridge no. 1 in which the predicted failure load of 471 kN compares reasonably well with the experimentally obtained value of 455 kN. The latter two predicted collapse mechanisms are shown in Fig. 3.14. Note also that in Fig. 3.14a the predicted diffused hinges are similar to those observed experimentally (shown in Fig. 3.10).

Further details of these analyses are provided elsewhere.[11,17] Slightly more refined analyses may also be performed. For example, a Boussinesq model may be assumed for the intensity of pressures beneath the load, and crushing of the masonry may also be included in the analysis.[21] Currently research is being focused towards also modelling the soil explicitly, within a combined masonry–soil computational limit analysis framework. As indicated previously, the backfill, if present, is very important and deserves more attention from arch bridge researchers.

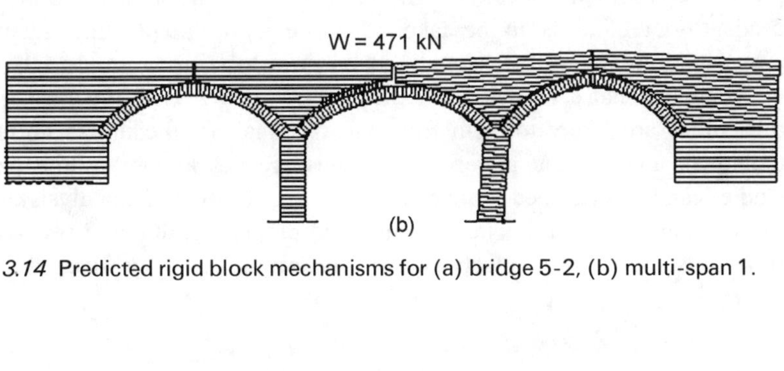

3.14 Predicted rigid block mechanisms for (a) bridge 5-2, (b) multi-span 1.

However, it must be pointed out that in the case of stretcher bonded arches, only deteriorated or 'shaken down' structures can be modelled with confidence using the rigid block analysis method (unless it is known that ring separation definitely will not occur). This is because the simple limit analysis formulation used is not appropriate when modelling the onset of ring separation, which involves quasi-brittle fracture processes. For this situation some sort of elastic analysis is required.

Elastic analysis

There are a number of types of problem for which the mechanism/rigid block analysis approaches described previously are clearly not well suited. Such problems include (i) determining in-service deflections and/or stresses; (ii) modelling elastic instability (e.g. snap-through); and (iii) modelling fracture (e.g. between rings in multi-ring brickwork arches).

Numerous elastic analysis techniques (e.g. the finite element method,[22] the discrete element method,[23] etc.) have been developed over the past few decades and most can in principle be applied to solve one or more of the above problems. Because of the plethora of methods available, basic principles rather than any particular method will be considered here.

In the context of masonry arch bridges two of the key issues associated with elastic analysis are: (i) in order to model hinge formation and subsequent collapse the analysis must be (materially) non-linear (note that this is in contrast to the rigid block analysis method, where a collapse analysis is a linear (optimisation) problem); and (ii) to properly model a phenomenon such as elastic snap-through the analysis must also be geometrically non-linear (i.e. the geometry from iteration n must be used for the calculations in iteration $n + 1$).

Many of the non-linear solution algorithms used in general-purpose finite element packages are not robust; hence convergence may be erratic and difficult to achieve. A simple non-linear solution procedure that is particularly suitable for masonry arches, and does not require the use of complex material models, was proposed by Castigliano. This procedure involves performing several linear elastic analyses and, after each, progressively removing areas of masonry containing tensile stresses until hinges and, eventually, a mechanism forms. This procedure is sometimes called a 'thinning elastic analysis' and has been widely used.[24,25] Care needs to be taken that the algorithm as implemented does properly allow hinges to change location prior to failure.

Assuming that a robust solution scheme is being used, practical modelling difficulties are likely to stem from the fact that: (i) the initial stress state is unknown, and (ii) the material properties are not known. While parametric studies can be performed to investigate the sensitivity of the analysis to changes in the boundary conditions and in material properties, this process can be very time-consuming, particularly when performing a fully three-dimensional analysis. This can make an elastic analysis of a masonry arch bridge analysis tedious to perform in practice.

Analysts should also be aware that when a non-zero value for the masonry tensile or shear strength is specified, serious modelling difficulties can arise. In 1976, Hillerbourg,[26] while attempting to model fracture in concrete, demonstrated that objective (mesh-size independent) results could only be obtained if suitable fracture parameters were included in the analysis (i.e. the fracture energy G_f, shown in Fig. 3.5, must be included in the analysis). This finding appears to have gone largely unnoticed in the masonry community, at least until recently. An ambitious programme of experimental work carried out in the Netherlands in the 1990s for the first time provided fracture parameters that could be used for modelling masonry.[27]

There are essentially two main approaches to the modelling of cracking in masonry: 'smeared crack' models and 'discrete crack' models. In general, 'discrete crack' models will provide much more reliable and realistic results, but these can be prohibitively computationally expensive when applied to large structures.

However, in the context of masonry arch bridges, while even basic material properties such as compressive and tensile strengths are unknown (and must be estimated using engineering judgement) it may appear to be incongruous that additional – more poorly understood and difficult to measure – parameters should also be required for analysis. Hence it is perhaps understandable that even now constitutive models that do not include these important additional parameters are used for masonry arch bridge analysis. However, because the results from these models will always be highly dependent on mesh size, any conclusions drawn from such studies must be treated with due scepticism.

3.4.7 Repair and maintenance

Providing that mortar materials similar to those used originally are employed, repointing is a simple, effective and comparatively inexpensive means of keeping masonry structures in an externally good condition. However, it is clearly not an effective repair technique for major defects such as ring separation or longitudinal cracking. In these cases consideration may be given to introducing metallic or FRP stitches or anchors. For the case of ring separation these could be oriented in a radial direction at regular intervals around the arch barrel.

Recently a proprietary system has been developed which uses very long anchors inserted from above or below into the arch barrel, providing reinforcement tangential to the arch at the quarter-span points. The system contains the grout injected around an inserted anchor in a fabric sock. It has recently been suggested that because this system is quick and easy to install, it provides an environmentally sensitive solution to the problem of strengthening a substandard bridge.[28] However, this of course assumes that the technique is effective. In fact, with this and other similar techniques, engineers should be aware that though the outward appearance of the structure may be unaltered, the character of the masonry arch may be permanently changed. Thus, for example, its ability to articulate freely with changing environmental conditions may be permanently impaired. Additionally, although in the case of the system described, several short-term proving tests have been performed,[29] the long-term response of structures so repaired remains unknown. Currently the technique cannot be independently checked by third parties, which has to date limited its use on the railway network.

More traditional techniques of strengthening masonry arches include replacing existing backfill with concrete and 'saddling' an arch using a curved reinforced concrete slab placed directly on the arch extrados. Providing suitably weak concrete fill is employed, the former is probably preferable since the arch continues to act in the manner originally intended.

3.5 Metal and concrete bridges

3.5.1 Metal bridges

Clearly, existing metal railway bridges take a diverse range of forms, ranging from the impressive or historically important (e.g. Brunel's Saltash Bridge, the Forth Rail Bridge, Stephenson's remaining tubular bridge at Conway, etc.) through to the numerous workmanlike short- and medium-span beam bridges distributed all over the network.

To give an indication of the numbers of metal bridges of different types on the network, of a sample of 2236 overline rail bridges assessed as part of the 'BridgeGuard' programme between 1995 and 2000, 16% were metal girders (including half through girders), 5% were of jack arch construction, 2% comprised trough decks, 1% cast iron beams and 0.4% were metal trusses; the majority of the remaining bridges assessed were masonry arches, with some concrete bridges also.

In the case of the numerous beam bridges on the network, these could be designed using scientifically based methods from when they were first constructed (though fatigue was initially not well understood). Stresses in beams subjected to given working loads could be calculated and checked against suitably factored maximum permissible values.

For bridges which pushed the boundaries, additional studies were sometimes required. For example, in the case of Robert Stephenson's large tubular rail bridges at Conway and the Menai Straits, prototype models were built and proving tests performed.[30] These identified thin plate buckling in the compression zones of the tubular section as a failure mode (very different from the tensile zone failures which had been encountered previously, when cast iron beams were tested). Thus suitable stiffeners were added to reinforce the compression zone.

In the case of more conventional beam bridges, the main problems which arose were often due to poor detailing (e.g. bridges often had inadequate and/or unmaintainable connections between the main beams and the deck).

By the second half of the twentieth century welding had superseded riveting as the most common connection method. This period also saw the introduction across the network of a number of standard underline bridge designs for short and medium spans (initially approximately 6 m to 30 m). The original standard designs used steel plate girder main beams with steel cross-beams embedded in reinforced concrete floor slabs (e.g. Fig. 3.15). Since the initial standard designs were produced, other standard solutions have also been developed (e.g. incorporating box girder beam sections).

Clearly, standardisation reduced initial design and construction costs and can now potentially reduce assessment costs, since the need for detailed investigations merely to identify construction details on a bridge-by-bridge basis should be obviated.

However, unfortunately most metal bridges on the UK network predate standardisation, with the majority dating from between 1880 and 1920;[1] many are considerably older. Assessment of these very old structures can be quite difficult. For example, many existing overline spans on the railways comprise jack arches spanning between longitudinal metal beams, but the structural performance of this system is still not particularly well understood. The same is also true for lattice girder station footbridges, which appear to have performed better in practice than is suggested by simple structural analysis.

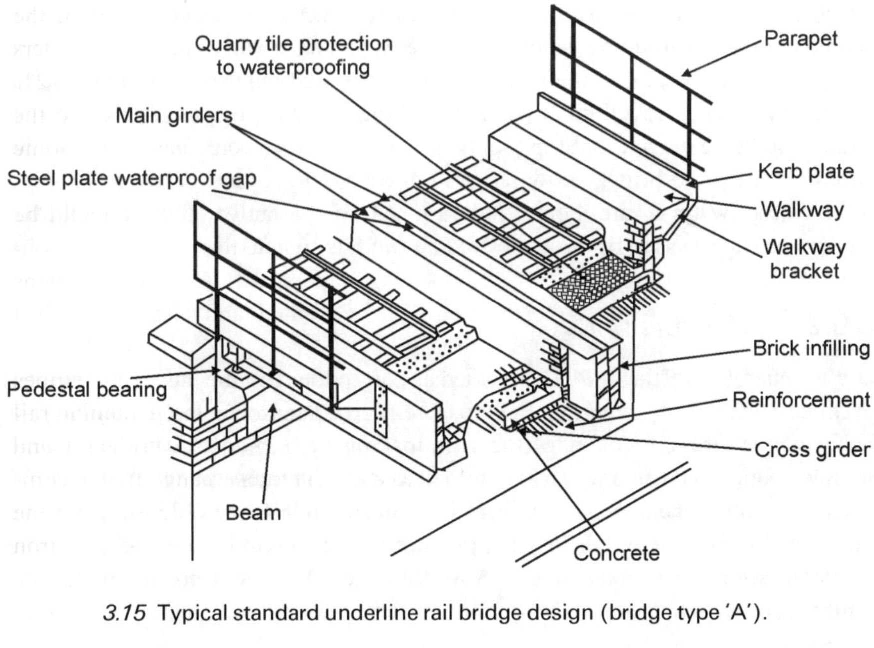

3.15 Typical standard underline rail bridge design (bridge type 'A').

3.16 Cahir Viaduct (1852), Ireland, which partially collapsed while carrying a freight train, 2003 (photo: Aidan Brosnan).

Despite gaps in our knowledge and understanding of old structures, fortunately failures are comparatively rare. However, this must not lead to complacency, and also it is of paramount importance that when failures inevitably do occur, these should be carefully analysed and lessons learnt. For example, derailment of freight wagons led to a spectacular failure in 2003 of an 1852 bridge at Cahir in the Republic of Ireland. Here derailed laden freight wagons impacted on and subsequently dislodged transverse beams over a large part of the bridge span (Fig. 3.16). While it was initially speculated that the derailment may have been caused simply by spreading of timber way beams which had become rotten (way beams span longitudinally over the transverse beams and should provide continuous support for the rails), an inquiry identified a more complex cause: dynamic interaction between the particular rigid wagons being pulled at the time, travelling at 40 mph, and the variable-stiffness bridge beams, which allowed a wagon to jump the tracks. A specific outcome has been that the speed limit for the bridge (now repaired at a cost of €2.6m) has been reduced when these wagons are being pulled.[31] More generally, it is hoped that awareness of the issue of dynamic interaction has increased.

3.5.2 Concrete bridges

Of the concrete overline bridges assessed as part of the 'BridgeGuard' programme between 1995 to 2000, about half were of reinforced concrete, the remaining half being of pre-stressed concrete (together making up 13% of all bridges in the sample). Reinforced concrete was heralded as a largely maintenance-free material when first introduced. However, there have been durability problems, as will be mentioned in Section 3.5.4. Precast, pre-stressed, concrete bridge decks became available after the Second World War. Because these were precast in factory rather than site conditions, the quality of the resulting construction was often excellent (though some early beams did have inadequate shear reinforcement). Post-tensioned beams have been more problematic, primarily due to problems in grouting the tendons on site (any voids left have allowed corrosion).

3.5.3 Fatigue in metal bridges

The fact that subjecting a piece of metal to many cycles can produce failure at a much lower load than would be required for static failure has been recognised by engineers since the mid-nineteenth century. At that time, as evidence of the phenomenon grew, the Board of Trade in the UK imposed more stringent limits on maximum permissible stresses (e.g. 5 tons/in^2 for wrought iron).[30]

Obviously, understanding of metal fatigue has grown very significantly since then. It is now known that the fatigue process is the result of several effects operating in sequence: initiation of a microscopic defect, slow incremental crack propagation and final unstable fracture.

In practical terms, modern bridge designers normally try to design out fatigue by ensuring that the stress range in each loading cycle is not too large. Additionally, welded connections are particularly prone to fatigue failures, and weld details need to be carefully specified. However, many early metallic bridges have design details which do not meet current fatigue criteria, and the severe consequences of failure mean that these continue to represent a risk.

3.5.4 Corrosion

In addition to fatigue, corrosion is the other main problem affecting metallic bridges. Of the materials used on the rail network, cast iron is the most resistant to corrosion, followed by wrought iron and then steel, the least resistant material. Certain areas of bridges are particularly vulnerable, for example the bottom flanges of plate girders or at footway level in the case of overline bridges with large plate girder edge beams.

In general, corrosion can be prevented by using an effective paint system. Though there will be situations when it is not cost-effective to paint a bridge if the latter is known to have a limited remaining lifespan, unfortunately rusting bridges look unsightly and this strategy tends not to be appreciated by the general public!

In the case of concrete bridges, corrosion is also a significant problem. In early concrete bridges the reinforcement was often not adequately protected by concrete cover. Additionally, the composition of the concrete was often undesirable (e.g. high water : cement ratio, use of unwashed marine aggregates, etc.). De-icing salts applied in cold weather to the road surface of overline bridges are a particularly common cause of corrosion in steel reinforcement.

Many techniques have been developed for repairing concrete bridges but unfortunately most are labour intensive, time-consuming and hence relatively expensive.

3.5.5 Assessment of metal and concrete bridges

In 2001 a new limit-state assessment code of practice was introduced for UK railway bridge assessment,[32] superseding the permissible stress approaches used since the emergence of the railways (except in the case of cast iron bridges). Limit-state approaches are already commonly used in the design of new structures; these properly distinguish between ultimate (collapse) and serviceability conditions (i.e. 'limit states'), and rationally allocate partial safety factors according to the degree of uncertainty associated with each individual parameter (e.g. dead load, material strength, etc.). This represents a very significant step in the modernisation of the assessment process for UK rail bridges, albeit at the expense of slightly increased complexity and hence burden on assessment engineers.

3.6 Parapets

Following high-profile cases of road vehicle incursions onto railway tracks (most notably the incident in 2001 at Great Heck near Selby), public interest in the efficacy of bridge parapets and approach restraint systems has recently increased significantly.

As with other structures on the railways, there are a huge variety of parapets currently in use. In addition to modern 'high containment' parapets constructed using steel or reinforced concrete, masonry parapets are also common. Assuming a given parapet has escaped reconstruction following impact-induced damage, this might typically be around 150 years old.

Prior to the relatively recent introduction of standard designs, parapets were designed individually, the parapet often melding seamlessly with the rest of the structure. Metal overline bridge parapets range from those comprising ornamental cast ironwork to utilitarian deep plate girder edge beams which also support the traffic loading.

Recently the performance of masonry parapets in particular has become an active research area. Because of their ubiquity, this section will focus on recent research findings and identify future research needs.

3.6.1 Masonry parapets

Masonry parapets were almost always constructed on masonry arch bridges and were also commonly constructed on wrought iron and other beam bridges.

Masonry parapets may be subject to several forms of loading. For example, underline bridge parapets are frequently required to carry cables and other services. Unfortunately the latter are often concentrated on only one side of the parapet, sometimes leading to overturning failure, either while additional cables are being placed or during subsequent high winds. However, the remedies are simple: remove redundant cabling; realign cables to reduce the overhang; in the last resort consider use of retrofitted anchor bars to tie down the wall to the bridge deck or spandrel wall below. However, another loading regime for overline rail bridge parapets which must be considered and which is altogether more complex to analyse is that due to vehicle impact.

Though anecdotal evidence has often indicated that masonry parapets may have the ability to contain vehicles, until recently there was no adequate means of verifying this. A problem has been that while modern steel and reinforced concrete parapets are commonly designed to resist a static lateral load which is deemed to be equivalent to the envisaged dynamic impact load, when a masonry parapet of reasonable dimensions is checked using the same 'equivalent static load' approach it will invariably be found to fail.

However, replacing all masonry parapets with more modern alternatives presents several problems: (i) the replacement may be aesthetically unacceptable; (ii) installing a replacement may be technically difficult since anchoring it to the spandrel wall (or ironwork) below is likely to be problematic; and (iii) replacing large numbers of masonry parapets is likely to be very costly.

Faced with the choice of either replacing all existing masonry parapets with modern alternatives in due course (with an estimated bill running into billions of pounds) or funding research to demonstrate that existing masonry parapets do have some containment capacity, in the 1990s local authorities in the UK chose the latter. Their research programme was facilitated by the County Surveyors' Society (CSS). To be deemed successful, an errant vehicle should not penetrate through the parapet, and should also be deflected by an acceptably small amount, so as not to come into the path of oncoming traffic.

The research project comprised actual vehicle tests (Fig. 3.17), some materials characterisation testing and finite element modelling studies. For the latter, artificially high values for the unit mortar shear and tensile strengths had

3.17 Aftermath of car impact test on a masonry parapet (photo: Matthew Gilbert).

to be used to provide good correlation with the wall tests. The research led eventually to the production of a British Standard covering masonry parapets.[33] This contains 'lookup' charts allowing the wall thickness required to contain a given speed of car to be determined. However, no indication is given as to the likely amount of masonry that might be ejected in an impact event. For obvious reasons this may be very important in the case of rail overline bridge parapets.

Though clearly having the advantage of realism, when performing impact tests using real vehicles it can be difficult to isolate masonry response from vehicle response. Hence more controllable laboratory tests on walls were also performed following the County Surveyors' Society research programme. Initial work comprised 21 full-size walls,[34] tested using a purpose-built testing rig, with the main findings summarised below:

- Mortar-bonded walls resist impact loading in two phases: initial elastic action prior to the formation of fracture lines, followed by gross displacements, resisted by friction at the base and both out-of-plane and in-plane inertial forces.
- Short walls and walls impacted close to their ends are significantly less able to resist impact loadings.
- Finite element modelling of walls using a discrete modelling strategy can be used to identify the failure mechanisms encountered. Using this approach, both weakly bonded and mortared walls can be modelled.
- A simplified mechanism-based method of dynamic analysis[35] can provide reasonable predictions of the behaviour of mortared walls.

However, there were still unanswered questions about the dynamic material response and also about how best to strengthen inadequate masonry parapets. Thus a follow-up research programme was instigated which comprised tests on laboratory walls with added reinforcement, together with associated modelling and materials characterisation studies.[36] The conclusions of this research were as follows:

- Large panel rather than loose block failure modes will only occur in an unreinforced wall if the unit-mortar bond strength is moderately high (e.g. equivalent to at least 1:1:6 cement:lime:sand mortar, when used with low absorbency engineering bricks).
- Use of retrofitted diagonal bar reinforcement is likely to be successful in preventing loose block failure modes, even when very weak mortar is used. Conversely, the use of bed-joint reinforcement alone is likely to be ineffective.
- Providing that masonry joints are modelled in a suitably detailed way (i.e. including joint fracture energy and joint dilatancy), there is no need to use artificially high 'dynamic' material properties in numerical models in order to achieve good correlation with the wall test results.

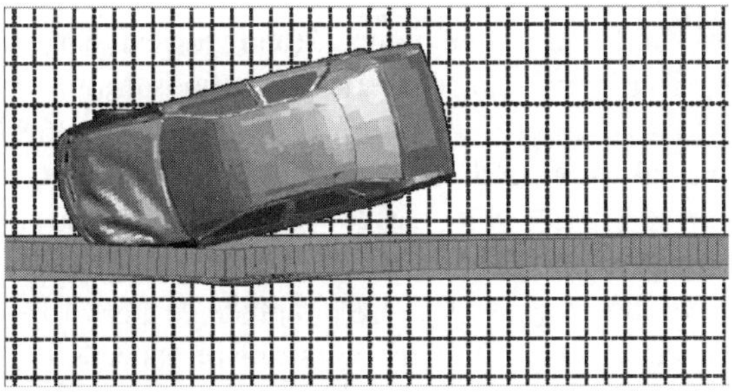

3.18 Numerical model of a brickwork wall subject to a 70 mph car impact (source: Hobbs, B., Gilbert, M., Molyneaux, T.C.K., Newton, P., Beattie, G. and Burnett, S.J., 'Improving the impact resistance of masonry parapet walls', *Proceedings of the Institution of Civil Engineers, Structures & Buildings*, Vol. 162, pp. 57–67, 2009.

Figure 3.18 shows a typical plot of a detailed numerical model of a car impact on a masonry wall.[36]

To conclude, from the situation a decade ago where very little was known about the performance of masonry parapets, our level of understanding is now much greater. We also now have effective masonry parapet strengthening methods at our disposal.

There are of course many remaining issues: masonry parapets, reinforced or otherwise, will seldom be capable of containing heavy goods vehicles. Furthermore, since it will often be unwise to anchor retrofitted wall reinforcement into an underlying old bridge deck or spandrel wall, a substantial impact from a heavy goods vehicle could in some cases cause complete overturning of a wall retrofitted with reinforcement. This might actually be a worse scenario than if the wall were not reinforced at all, since in the latter case only fragments of masonry would be likely to be ejected. Methods of restraining the ends of such reinforced walls therefore need to be developed.

3.7 Future trends

The recent introduction of limit-state methodology for the assessment of most UK railway bridge types represents a valuable step in modernising the way existing structures are assessed on the rail network. However, there are still anomalies in the current assessment procedures. For example, the semi-empirical MEXE method of assessment for masonry arch bridges has a dubious basis and hence cannot be considered to be a rational analysis tool. Unfortunately, at present a barrier to phasing out MEXE is our current inadequate understanding of phenomena such as arch soil–structure interaction, which means that current mechanism analysis programs (the likely natural successors to the MEXE method for initial assessment) are prone to give conservative predictions of bridge strength. It is therefore hoped that research currently being undertaken will soon lead to the development (and subsequent widespread application) of improved analysis and assessment methods for masonry arch bridges.

Additionally, although already used on certain critical structures on the rail network, instrumentation of bridges with sensors (e.g. crack gauges) is likely to become increasingly common, with the sensors being networked to central control units. This means that engineers can be rapidly deployed to inspect a given structure when preset limits (e.g. crack widths) are exceeded.

Non-destructive testing (NDT) methods are also likely to continue to improve, and to be used increasingly in routine bridge inspections, for example to identify hidden voids in masonry arch bridges. However, despite advances in NDT technology, for the foreseeable future it is likely to remain impossible, for example, to accurately determine the spatial variation of the bond integrity in multi-ring brickwork arch bridges. This means that it is also currently impossible to accurately compute the ultimate strength of such bridges. While this is only a problem if the computed carrying capacity proves to be inadequate when 'worst case' assumptions about the integrity of the inter-ring joint have been made, if this is the case then an assessment engineer currently faces a real quandary. In the future it is likely that probabilistic assessment methods will be used in such cases (i.e. the assessment input parameters are entered as ranges, rather than as discrete values, with the output being in the form of a probability distribution). Additionally, a greater understanding of the performance of masonry under cyclic loading should allow the residual life of existing structures to be estimated more reliably.

More generally, developing rational management strategies to encompass the huge variety of bridges on the UK rail network is certainly difficult. Currently, Network Rail appears to be moving away from the minimal intervention 'reactive' approach to infrastructure of their predecessor Railtrack, towards a whole-life-cycle planned maintenance approach. It is to be hoped that this continues into the future.

3.8 Sources of further information

Partly due to their ubiquity, this chapter has focused particularly on the performance of masonry arch bridges. Unfortunately, because this structural form has been neglected by structural engineers and researchers for much of the last century, there are even now relatively few useful books on the engineering aspects of masonry arch bridges.

Of the books which do exist, Heyman's 1982 volume *The Masonry Arch*[19] presents a well-argued case to support the application of mechanism analysis to masonry arches. John Page's text *Masonry Arch Bridges*,[9] published in 1993, although rather brief, does contain a fairly comprehensive list of references for the interested reader. Sowden has edited a useful volume for those interested in the maintenance, repair and rehabilitation of masonry arches.[37] More recently, CIRIA have published a useful volume on condition appraisal and remedial treatment of masonry arches.[38]

Scientific interest in masonry arch bridges appears to be growing in continental Europe particularly, and the proceedings of the series of 'International Arch Bridges Conferences' provide useful information on recent research.[39–43]

Useful general books on dealing with old bridges include a Highways Agency sponsored volume on the conservation of bridges.[2]

3.9 Conclusions

A very large proportion of the bridges on Europe's rail networks are of masonry arch construction, with most of these being well in excess of 100 years old. Despite the fact that the masonry arch is an ancient form, their basic behaviour is still inadequately understood. Fortunately, there have been few recorded instances of arch bridges failing without warning, indicating that frequent inspection remains a fall-back position while improved analysis and assessment methods are being developed. Considering fatigue failure, it has been demonstrated that certain specific types of arch bridge (e.g. those containing multi-ring brickwork barrels) can exhibit marked reductions in ultimate carrying capacity after a large number of loading cycles have been applied. However, further investigations are required in order to further enhance our understanding of fatigue failure in masonry structures.

Compared with masonry arch bridges, the basic structural behaviour of bridges of other types is generally much better understood, although bridge assessment and management can be challenging nonetheless. As regards the specific issue of fatigue failure, while the designers of modern metallic bridges usually try to design out fatigue (by ensuring that the stress range in each loading cycle is not too large), many early metallic bridges on the network have design details which do not meet current fatigue criteria. Such bridges warrant careful assessment.

3.10 References

1. Network Rail, *Technical Plan, Section 9: Plans by Asset Type*, London, 2003.
2. Tilly G, *Conservation of Bridges*, London, Spon, 2002.
3. Salmon E H, *Materials and Structures, Vol 2*, London, Longmans, 1938.
4. Brunel I, *The Life of Isambard Kingdom Brunel, Civil Engineer*, London, Longmans, 1870.
5. Owen J B B, 'Arch bridges', in *The Works of Isambard Kingdom Brunel*, ed Pugsley A, ICE/University of Bristol, 1976.
6. *New Civil Engineer*, 19 June 1992, 5.
7. Barlow W H, 'On the existence (practically) of the line of equal horizontal thrust,

and the mode of determining it by geometrical construction', *Min Instn Civ Engrs*, 1846, **5**, 162–182.
8. Ruddock E C, 'Hollow spandrels in arch bridges: a historical study', *The Structural Engineer,* 1974, **52**(8), 281–293.
9. Page J, *Masonry Arch Bridges*, London, HMSO, 1993.
10. Page J, *Load test to collapse on the three span brick masonry arch Rotherham Road Railway Bridge*, TRL Project Report, PR/CE/49/94 (unpublished), 1994.
11. Melbourne C and Gilbert M, 'The behaviour of multi-ring brickwork arch bridges', *The Structural Engineer*, 1995, **73**(3), 39–47.
12. Melbourne C and Hodgson J A, 'The behaviour of skewed brickwork arch bridges', *Third Int Conf Bridge Management*, Guildford, 1996.
13. Hughes T G, Davies M C R and Taunton P R, 'The small scale modelling of masonry arch bridges using a centrifuge', *Proc Inst Civ Engrs, Structures and Buildings*, 1998, **128**, 49–58.
14. Hughes T G, Roberts T M, Goutis G and Bell B, 'Serviceabilty of masonry arch bridges', *STRUMAS VI*, Rome, 2003, 126–135.
15. Melbourne C and Aluaimi N M, 'Behaviour of multi-ring brickwork arches subjected to cyclic loading', *Third Int Conf Arch Bridges*, Paris, 2001.
16. Melbourne C and Tomor A, 'Fatigue performance of composite and radial-pin reinforcement in multi-ring masonry arches', *Fourth Int Conf Arch Bridges*, Barcelona, 2004, 427–433.
17. Melbourne C, Gilbert M and Wagstaff M, 'The collapse behaviour of multi-span brickwork arch bridges', *The Structural Engineer*, 1997, **75**(17), 297–305.
18. Pippard A J S and Baker J F, *The Analysis of Engineering Structures*, London, Arnold, 1943.
19. Heyman J, *The Masonry Arch*, Chichester, Ellis Horwood, 1982.
20. Livesley R K, 'Limit analysis of structures formed from rigid blocks', *Int J Num Meth Eng*, 1978, **12**, 1853–1871.
21. Gilbert M, 'RING: a 2D rigid-block analysis program for masonry arch bridges', *Third Int Conf Arch Bridges*, Paris, 2001.
22. Bathe K J, *Finite Element Procedures*, Englewood Cliffs, NJ, Prentice Hall, 1996.
23. Cundall P A and Strack O D L, 'A discrete numerical model for granular assemblies', *Geotechnique*, 1979, **29**, 47–65.
24. Choo B S, Coutie M G and Gong N G, 'Finite element analysis of masonry arch bridges using tapered beam elements', *Proc Instn Civ Engrs*, 1991, Part 2, **91**, 755–770.
25. Bridle R J and Hughes T G, 'An energy method for arch bridge analysis', *Proc Instn Civ Engrs*, 1990, Part 2, **89**, 375–385.
26. Hillerbourg A, 'Analysis of crack formation and crack growth in concrete by means of fracture mechanics and finite elements', *Cement and Concrete Research*, 1976, **6**, 773–782.
27. Rots J G, *Structural Masonry (CUR Report 171)*, Rotterdam, Balkema, 1997.
28. Steele K, Cole G, Parke G, Clarke B and Harding J, 'Environmental impact of brick arch management', *Proc Inst Civ Engrs, Structures and Buildings*, 2003, **156**(SB3), 273–282.
29. Sumon S K, 'New reinforcing systems for masonry arch rail bridges', *Second Int Conf Railway Engineering*, London, Engineering Technics Press, 1999.
30. Timoshenko S P, *History of Strength of Materials*, New York, Dover, 1983.
31. Irish Railway Record Society, 'Cahir viaduct', *Journal of the Irish Railway Record Society*, 2005, **155**.

32. Network Rail, *The Structural Assessment of Underbridges*, Company Code of Practice RT/CE/C/025, 2004.
33. BS6779, *Highway Parapets for Bridges and Other Structures, Part 4: Specification for Parapets of Reinforced and Unreinforced Construction*, London, BSI, 1999.
34. Gilbert M, Hobbs B and Molyneaux T C K, 'The performance of unreinforced masonry walls subjected to low-velocity impacts: experiments', *Int J Impact Eng*, 2002, **27**, 231–251.
35. Gilbert M, Hobbs B and Molyneaux T C K, 'The performance of unreinforced masonry walls subjected to low-velocity impacts: mechanism analysis', *Int J Impact Eng*, 2002, **27**, 253–275.
36. Beattie G, 'Joint fracture in reinforced and unreinforced masonry under quasi-static and dynamic loading', PhD thesis, University of Liverpool, 2003.
37. Sowden A M (ed), *The Maintenance of Brick and Stone Masonry Structures*, London, Chapman and Hall, 1990.
38. McKibbins L, Melbourne C, Sawar N and Sicilia Gaillard C, *Masonry arch bridges: condition appraisal and remedial treatment*, Report C656, CIRIA, 2006.
39. Melbourne C (ed), 'Arch bridges', *First Int Conf Arch Bridges*, Bolton, Thomas Telford, 1995.
40. Sinopoli A (ed), 'Arch bridges: history, analysis, assessment, maintenance and repair', *Second Int Conf Arch Bridges*, Venice, Balkema, 1998.
41. Abdunur C (ed), 'Arch01', *Third Int Conf Arch Bridges*, Paris, ENPC, 2001.
42. Roca P and Molins C (ed), 'Arch bridges IV', *Fourth Int Conf Arch Bridges*, Barcelona, CIMNE, 2004.
43. Lourenco, P B, Oliveira D V and Portela A (ed), 'Arch07', *Fifth Int Conf Arch Bridges*, Madeira, University of Minho Press, 2007.

4

Safety and reliability issues affecting escalators and moving walkways in railway stations

K. BEHRENS, formerly of ThyssenKrupp, Germany

Abstract: This chapter reviews the safety and reliability of escalators and moving walkways in railway stations. It discusses safety issues such as transition curvature for escalators and reversal of travel systems as well as features such as balustrades, skirt deflectors and handrails. It also reviews issues affecting service life such as misuse and vandalism.

Key words: railway escalators, moving walkways, transition curvature, reversal of travel systems

4.1 Introduction

Today escalators and moving walkways are part of our everyday life and it is not possible to imagine railway and underground stations without them. However, they are powerful and potentially dangerous machines. The growing number of escalators and moving walkways installed in Europe after the Second World War required the drawing up of guidelines on safe design and operation, particularly as not all European countries had their own standards or national regulations for this type of conveyance. Escalators and moving walkways fall under the category of miscellaneous construction products which need to conform to the European Union CE-marking (89/392/EEC).[1] Today the European standard EN 115[2] provides specific guidelines for the safe construction and installation of escalators and moving walkways.

Escalators and moving walkways in railway installations (see, e.g., Fig. 4.1) require a special approach because, in contrast to escalators and moving walkways in buildings such as banks, office buildings and department stores, they are exposed to completely different conditions. These include:

- Exposure to the weather (e.g. rain, snow and ice)
- Not being in air-conditioned surroundings
- Periods of operation over 140 h/week
- Being in relatively open but isolated areas (increasing the risk of vandalism, for example).

These special operating conditions need to be balanced with typical expected lifespans of at least 20 years.

4.1 Typical railway station escalator layout.

4.2 Safety issues affecting escalators and moving walkways

Current guidelines require that an escalator or moving walkway should stop automatically in the following circumstances:

- Absence of control voltage
- Fault to earth of a circuit
- Overload
- Operation of the control devices at overspeed or unintentional reversal of the direction of travel
- Operation of the auxiliary brake
- Breakage or undue elongation of parts immediately driving the steps, pallets or belt, e.g. chains or racks
- (Unintended) reduction of the distance between the driving and return devices
- Foreign bodies being trapped at the point where the step, pallets or belt enter the comb
- Stopping of a succeeding escalator or moving walkway where an intermediate exit does not exist
- Operation of the handrail entry guard
- Sag in any part of the step or pallet so that meshing of the combs is no longer ensured
- Operation of the control devices to detect a missing step or pallet
- Operation of the handrail speed monitoring devices caused by a broken handrail.

A device should be provided to detect the lifting of the braking system and all safety contacts or safety circuits must accord with EN 115.[2]

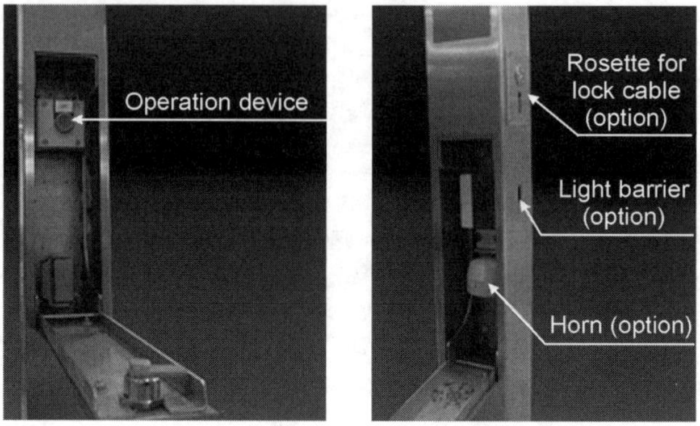

4.2 Layout of escalator emergency stop devices.

Emergency stop devices should be placed in conspicuous and easily accessible positions at or near the places where an escalator or moving walkway starts and stops. For high-rise escalators above 12 m or moving walkways with a length of more than 40 m, additional emergency stop switches must be provided. Pull switches located in highly visible positions are normally used for public service escalators or moving walkways. These pull switches should be partly mounted on the walls or on special columns at a height of approximately 170 cm. These columns may also house key-operated switches for use by staff to protect against improper use. Stop switches must conform to EN 418. The switch must work according to the principle of positive opening operation. The emergency stop switch must be red. The background behind the switch should be yellow if feasible. Figure 4.2 shows an example of the layout of emergency stop devices.

4.2.1 Transition curvature for escalators

The radii of curvature at the upper and lower transitions are parameters that may vary for public service escalators. In the transition areas between the inclined and horizontal part of the escalator, the user experiences acceleration forces which must be compensated by weight changes. It is recommended that public service escalators for speeds above 0.65 m/s have a lower transition of 2 m and an upper transition of 2.60 m. The revised EN 115[2] has restated that the radii be dependent on inclination and travel speed. Acceleration forces in the area of the transition curvature are:

- Radial acceleration: $a_R = V^2/r = 0.5^2/1 = 0.25\,\text{m/s}$
- Horizontal acceleration: $a_H = a_R \sin 35^\circ = 0.143\,\text{m/s}$
- Vertical acceleration: $a_V = a_R \cos 35^\circ = 0.204\,\text{m/s}$

These are illustrated in Fig. 4.3.

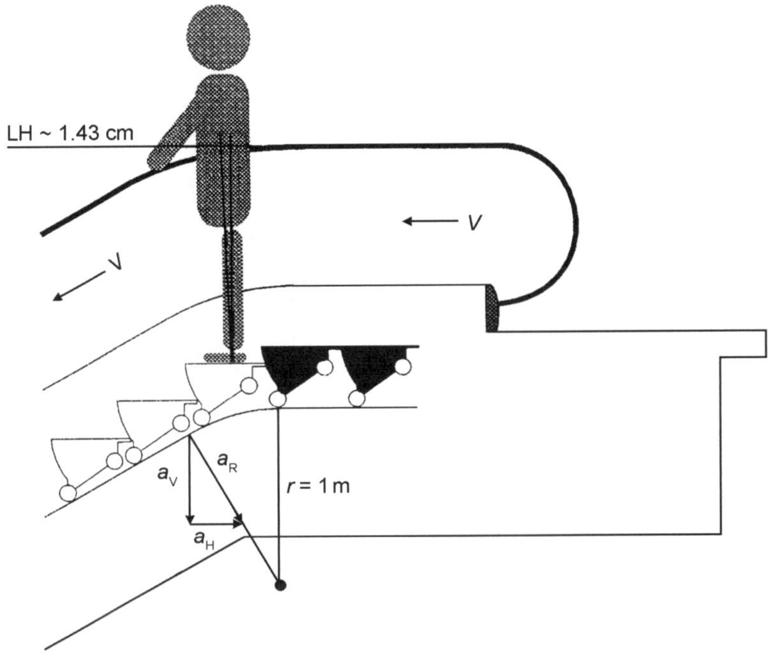

4.3 Acceleration factors affecting transition curvature.

4.2.2 Balustrades

Normally, climbing onto the balustrade (Fig. 4.4) is possible only at the lower and upper landings. Particularly in the case of glass-enclosed balustrades (Fig. 4.5), there have been cases where children have incurred injury through playing on the newels or riding on the balustrades. This must be taken into account when

4.4 Solid balustrades.

4.5 Glass balustrades.

designing connector railings to the escalator, e.g. by raising the height of the connector railings. On high-deck balustrades, anti-slide devices should be provided. For ergonomic reasons a balustrade height of 1.0 m is preferable, though a height of 1.10 m may be required. An angle of inclination of at least 25° should prevent children climbing on the balustrade and riding on the handrail. A survey by the Universitat Politècnica de Catalunya in Spain suggests the inclination should be at least 27°.

4.2.3 Comb teeth

The angle of the comb teeth (Fig. 4.6) requires particular attention when children's buggies, luggage, shopping trolleys, etc., are likely to be used by passengers. In these circumstances the angle should not exceed 19°.

4.6 Comb plate.

4.7 Skirt brushes.

4.2.4 Skirt deflector

All escalators in the UK since 1995 have been fitted with skirt brushes (Fig. 4.7). More than 10 years' experience with skirt deflector devices in the UK prove that the number of accidents has been reduced significantly. The installation of skirt brushes was adopted by EN 115[2] as a recommendation in November 2000 and published as amendment EN 115[2] in 2005.

4.2.5 Handrails

The handrail speed tolerance should be between 0% and 2%, otherwise stumbles may be caused to passengers being imperceptibly dragged in a reverse direction. The revision of EN 115[2] states that a 15% deviation from the nominal speed should stop the escalator/passenger conveyor after 15 s.

4.2.6 Escalators as emergency exits

The use of out-of-service escalators in countries such as Germany is illegal. In these circumstances, out-of-service escalators can only be used in underground railway stations with special permission.

4.2.7 Starting mechanisms

In general, escalators and moving walkways should not be started when passengers are on the steps or pallets. Escalators with high travel heights that have been stopped by use of the emergency switch should only be started again by authorised persons. The restart acceleration must not exceed 0.5 m/s.

4.2.8 Reversal of travel systems

Reversal of travel direction (alternating operation) is being increasingly used at poorly frequented underground railway stations. This option can be used with an automatic restart operation which ensures that nobody is using the escalator or passenger conveyor when the travel direction is being changed. These systems use a traffic-light signal system with green or blue (with arrow) and red (with white bar) (Fig. 4.8). An additional light indicating 'oncoming traffic' is used at the top and bottom entry and landing areas.

4.2.9 Supporting structure

For public service escalators and moving walkways, the maximum calculated or measured deflection should not exceed 1/1000 of the distance between supports, based on the static load of 5000 N/m.

The design of the supporting structure should be subject to a resonance test, in addition to deflection testing in the case of slim supporting structures or those with a poor dampening effect. The resonance frequency of human footsteps is between 1.5 and 2.8 Hz.

4.2.10 Lighting

There is no automatic requirement for the installation of comb lighting. Appropriate illumination can be delivered by general lighting as set out in standard operating manuals. On indoor escalator or moving walkways lighting should not be less than 50 lx at the landings and on outdoor escalators or moving walkways not less than 15 lx at the landings, measured at floor level.

4.8 Traffic light system for reversal of travel systems.

4.2.11 Stopping distances

Stopping distances must conform to European Standard EN 115.[2] The braking system must steadily decelerate the escalator or walkway. In a CEN risk assessment,[3] a maximum deceleration rate of 1 m/s was calculated so as not to cause passengers to stumble.

4.3 Reliability and service life issues affecting escalators and moving walkways

With most machines, it is advisable to conduct preventative maintenance. Escalators and moving walkways are no exception. After a set time interval the components should be reconditioned or replaced, irrespective of wear at this point in time. Step chains in particular should be monitored since they are particularly prone to wear.

4.3.1 Misuse of emergency stopping devices

For escalators, particularly public service escalators, potential misuse is an important issue. By far the most common occurrence is misuse of the emergency stop at the entry and exit areas of the escalator. Switching the escalator back on can take a long time, especially at stations where no staff are present. In order to reduce misuse of emergency stopping devices, in 1980 the escalator industry developed a switch that enables an automatic escalator restart. After a test phase this switch was adopted in 1991 by EN 115.[2] Availability of the escalators in public service systems was thereby significantly increased (to a level of over 99% in Germany where the automatic restart switch is used by virtually all large operators). This switch has become such a success that hardly an escalator in Germany is not equipped with this feature. Retrofitting of automatic restart switches to older systems is also possible.

4.3.2 Service life

Escalators and moving walkways in public service installations have a life expectancy of at least 20 years. The resulting 150 000+ operating hours, compared with the 2000–3000 of an automobile, show escalators and moving walkways to be extremely durable. The main components are expected to last over 20 years. These main components include:

- Supporting structure
- Primary shaft
- Drive unit
- Steps

Table 4.1 Corrosion protection systems

Corrosion protection system	Length of protection (in years)
Zinc galvanisation	20–25
Coating systems	10–15
Duplex system	>35

- Tensioning station
- Control unit.

Zinc galvanisation is a proven anti-corrosion measure for supporting structures in outdoor locations. Zinc corrodes far slower than steel and therefore prevents corrosion of the steel as long as an adequate coating of zinc is present. Depending on atmospheric conditions, the zinc coating degrades by 10 to 100 g/m. That means the average life expectancy of the protective zinc coating is limited to approximately 20 years, after which the unprotected structure is fully exposed to corrosive activity. A longer life expectancy can be achieved by an additional zinc coating (using e.g. a Duplex system), as indicated in Table 4.1.

Particular requirements include:

- Roller bearings: in consideration of the complex of loads acting on a component and general statics, the roller bearings should be designed for a life expectancy of 146 000 hours.
- Steps/pallets: the steps/pallets should be tested for at least 5×10^6 cycles, corresponding to European Standard EN 115.[2]
- Controls: the components controlling the electrical safety devices should be selected for continuous operation corresponding to EN 115.[2]
- Tensioning: the step/pallet chains should be tensioned continuously, a monitoring system continually checking the movements of $\pm$20 mm. For public service escalators and moving walkways, circulating chain wheels are preferable.

4.3.3 Vandalism

Vandalism can significantly reduce the life expectancy of an escalator. Damage by vandalism takes many forms: disfiguration of balustrade panels, cut rubber handrails, kicked-in handrail inlets, and graffiti on panels or outer cladding. Damaged or soiled escalators negatively influence the image of the railway as a preferred mode of transport. Vandalism is most common in cities with 20 000 to 100 000 inhabitants. In larger cities the use of escalators is far higher and there is less opportunity for acts of vandalism to go unnoticed. Experience in Berlin, Munich, Paris and Stockholm has shown that stainless steel panels with a fine 'satin finish' (Fig. 4.9) surface provide good protection and are easier to clean

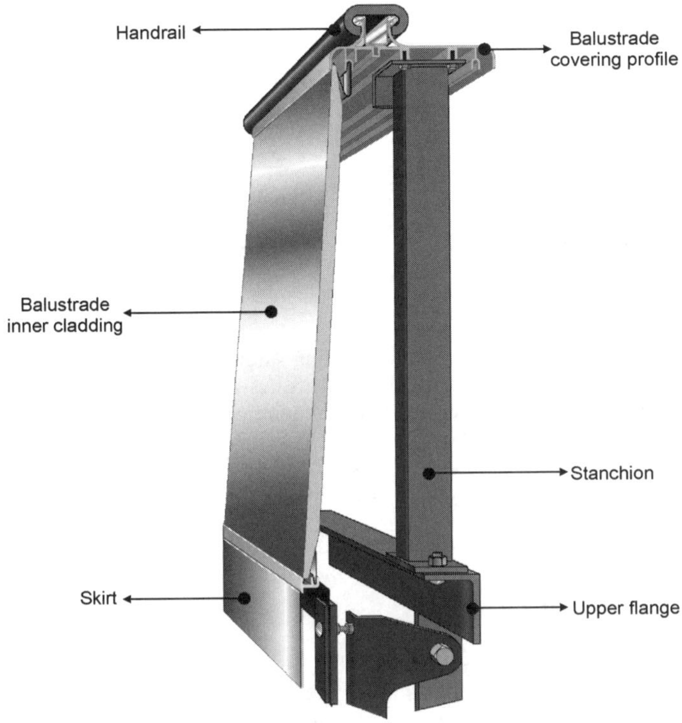

4.9 Stainless steel balustrade cross-section.

4.10 Pull switch.

than rough or painted surfaces. Control panels, stop switches or pull switches (Fig. 4.10) should be housed in stainless steel columns approximately 1.7 m high. Positioning at this height makes them more visible and therefore less easy to vandalise without being seen.

4.4 References

1. Directive on Machinery Safety (89/392/EEC).
2. EN 115: Safety rules for the construction and installation of escalators and moving walks.

5

Design, safety and reliability of lifts in railway stations

H.-P. KOHLBECKER, Deutsche Bahn Station and Service AG, Germany

Abstract: This chapter reviews design, safety and reliability issues affecting lifts in railway stations. It considers lift design, size and design specifications, vandalism resistance requirements, technical equipment and safety systems as well as lift control systems.

Key words: railway station lifts, lift control systems

5.1 Introduction

Figure 5.1 shows an example of a historic railway station forecourt. After over 100 years of existence, the function and the content of railway station buildings have changed in pace with technical development.

Depending on individual circumstances, railway companies are revitalising old railway stations to meet new needs. This involves the redesign of the railway station infrastructure to meet higher quality standards and more varied requirements including those of railway passengers, business meetings and commercial services.

The key objective is to identify in what way the needs of rail travellers and other users, including those with restricted mobility, can best be met so as to enable trouble-free use of railway stations for everyone.

5.1 Historic railway station forecourt.

In many respects accessibility in rail travel depends on the design of railway stations and the technical facilities installed on railway platforms. The design of the station environment for easy access by those with restricted mobility is an essential condition for the social integration of disabled and elderly people. The aim should be to ensure barrier-free passenger traffic from the railway station to the train and vice versa. To guarantee high-quality design, installation and functioning of reliable and cost-efficient lift systems and escalators in railway stations, the entire system for conveying passengers through the station should be described in an overall planning guide for the station. This chapter points out some important requirements that must be met by the lifts in railway stations.

5.2 Lift design, size and design specifications

There are well-established technical regulations for the construction and operation of lift systems. Lifts should be oriented to the main stream of passenger traffic and their barrier-free accessibility ensured. The load rating should be defined based on the expected volume of traffic. In particular, the transport of cleaning machines for the railway platforms should be taken into consideration in the sizing of lifts. In front of lifts, sufficient loading space measuring at least 1.50 m × 1.50 m should be left clear.

The calculation of the lift car size should be based on anticipated needs. For railway stations in the German railway network, a car size of at least 1.10 m × 2.10 m has been defined (or 1.10 m × 1.40 m in exceptional cases where structural restrictions apply). Provision should be made for use of the lifts by wheelchair users (Fig. 5.2) and the transport of children's prams, bicycles, stretchers and luggage trolleys. Passenger lifts must be fitted with emergency voice communication systems. Entrances and exits should be barrier-free. The weatherproof design of lifts that are installed either entirely or partly outdoors is also an important consideration.

5.3 Vandalism-resistant requirements for railway station lifts

Lift shafts and cars should be designed as transparent structures wherever possible as this enhances the feeling of passenger security and reduces damage by vandalism. The lower part of the car walls and walls of the lift shaft must be protected against damage. A rust-resistant steel floor with a non-slip covering is recommended. Handrails should not be fixed to the glass wall. Surfaces must be wear-resistant and robust with a low cleaning requirement. Inflammable materials should be avoided.

Technical details on avoiding damage by vandalism are included in the European standard EN 81–71. In the lift car, only screws that cannot be

5.2 Example of a handicapped-accessible lift.

loosened with conventional tools should be used where possible. The installation of video monitoring helps considerably to reduce damage by vandalism.

5.4 Technical equipment and safety of lift systems

This section provides some technical information on lift system equipment. This information refers to certain key aspects which are essential for the safe operation of lifts at or in railway stations. It is important to refer to the relevant technical standards (especially EN 81).

Thanks to the new drive technologies in traction sheave lifts, the installation of cable lifts without machine rooms is increasingly common at railway stations. The energy savings and the reduction in structural costs from the elimination of the machine room are important arguments in favour of their installation. For retrofitting of lifts, construction times tend to be shorter as a machine room is no longer required. Hydraulic lifts are often now favoured only for transporting very heavy loads.

In planning and construction, measures necessary for environmental protection must be taken into consideration. Only environmentally friendly products and processes should be used. For lifts in buildings, halogen-free cabling is becoming increasingly common.

5.3 Example of the design of a handicapped-accessible operating panel in a lift car.

Particular importance should be attached to corrosion protection. All materials exposed to the weather have to be treated appropriately as a precaution. Special importance is attached to the use of stainless steel. Rust-resistant steel with resistance to intercrystalline corrosion (chlorides, chlorine ions, urine and its decomposition products) should be used in lift cars. If luggage trolleys are used in the railway station, a protective skirting should be installed over the lower part of the lift car. The lift car ceiling should be designed to enable the trouble-free housing of lighting and ventilation. Dazzling of engine drivers by strong lift lighting in transparent lift cars must be avoided.

The operating panel should be designed to enable access by the handicapped (Fig. 5.3). Non-flammable large buttons with tactile marking are recommended.

Emergency voice communication using automatic self-dialling to an office manned round the clock is indispensable. Automatic voice announcements at each stopping floor for lifts travelling between more than one floor also help the visually impaired.

In the area of the lift shaft and the car doors, increased impact loads, caused by crowding or vandalism, have to be expected. These requirements should be taken into account in lift design. If the outer housing is on railway platforms, a protective skirt is essential when luggage and baggage trolleys are used. This protective skirting can be constructed from a stainless steel circular tube, or a concrete base should made available by the building owner. With regard to drive systems, the use of gearless frequency-controlled sheave drives is becoming more widespread. The operating speed for lifts travelling between more than two floors should be selected at around 1.00 m/s.

5.5 Lift control systems

A direction-insensitive one-button collective system or, in the case of more than two lift stops, a two-button collective system, should be selected for lift command and control. The lift control system should be generally designed as a self-diagnostic PLC. This improves fault analysis and enables detailed tracking of lift availability. Fault analysis quickly reveals whether technical defects have occurred or a fault has been caused by, for example, vandalism or misuse of the lift. This system can be used to develop a preventative maintenance strategy.

Standard requirements for lift control systems in railway stations include:

- Standardised bus and communications interface for connection to building automation systems
- Available connection for a log printer
- Remote diagnostics and remote control facility
- Time-coded fault memory (at least 100 faults)
- Self-diagnostics and fault display as plain text display
- High electromagnetic compatibility
- Simple on-site parameterisation of the control system
- Wear monitoring, e.g. based on trip counters, hours-run meters, motor hour meters and door cycle counters.

Intercom stations must be installed in the lift car and in the machine room. If necessary, other stations can be installed on the roof of the lift car and the shaft pit. In the case of a power cut, the function of the emergency call systems should be ensured over a defined period of time (at least 1 hour). At least the station in the lift car should be designed as a hands-free intercom unit for open duplex communication.

All passenger lifts that pass through more than one fire compartment and do not end outdoors or on railway platforms should be equipped with a dynamic control system in case of fire. This control system should guarantee that when the fire alarm is activated, depending on the actual fire situation, the lift travels to an individually defined safe stop without exposure to fire danger.

Naturally, considerable importance should be attached to the maintenance and repair of the lifts. Only prompt fault clearing and lift maintenance can increase availability and minimise vandalism. To guarantee high availability, a maximum delivery time for essential replacement parts should be agreed with the manufacturers.

Index

Index